3rd International Conference on Aeronautical, Aerospace and Mechanical Engineering (AAME 2020)

IOP Conference Series: Materials Science and Engineering
Volume 816

Sanya, China
16 - 18 February 2020

ISBN: 978-1-7138-2803-7
ISSN: 1757-8981

Printed from e-media with permission by:

Curran Associates, Inc.
57 Morehouse Lane
Red Hook, NY 12571

Some format issues inherent in the e-media version may also appear in this print version.

Institute of Physics
Dirac House, Temple Back
Bristol BS1 6BE UK

Phone: 44 1 17 929 7481
Fax: 44 1 17 920 0979

techtracking@iop.org

Additional copies of this publication are available from:

Curran Associates, Inc.
57 Morehouse Lane
Red Hook, NY 12571 USA
Phone: 845-758-0400
Fax: 845-758-2633
Email: curran@proceedings.com
Web: www.proceedings.com

TABLE OF CONTENTS

MECHANICAL DESIGN MANUFACTURING AND AUTOMATION

AVIATION SCIENCE AND ELECTRONIC COMMUNICATIONS

IOP Conf. Series: Materials Science and Engineering **816** (2020) 011001 doi:10.1088/1757-899X/816/1/011001

Preface

This proceeding contains the scholarly papers presented in the 3rd International Conference on Aeronautical, Aerospace and Mechanical Engineering (AAME 2020) held in February, 2020 in Bangkok, Thailand.

Submissions to the conference have been reviewed by members of the Technical Committee consisted of professors and senior academics from reputable international universities in Australia, Asia, Europe and South Africa. The papers in this proceeding have passed the vigorous review process and amended to the satisfaction of reviewers. The face-to-face presentations have also undergone critiques by expert audience in the conference. It is on this basis that the published papers in this proceeding are good references to future research investigations and analyses.

Papers in the conference can be categorised according to the themes of the conference: aeronautical, aerospace and mechanical engineering. It is worth to note that the purpose of categorisation with these themes is only to help readers to search for papers. The proceeding as a whole presents a consistent focus on aerodynamics, aerostructures, mechanical strength and space exploration technologies. The distribution of papers among these themes is close to balanced, i.e. 9 papers in aeronautical, 6 papers in aerospace and 8 papers in mechanical engineering.

Among the 9 papers in aeronautical engineering, structural integrity of aerostructures such as airframe and wing are the most common topics. Another common topic is aerodynamics on aeronautical components which extends into analysis of forces and loading. The novel analyses exhibit in these papers push the boundary of knowledge and enable better aircraft design and manufacture.

The 6 papers in aerospace engineering are mostly concerning satellite design, operations and communication. Of particular interest is a paper on aerodynamics performance of hypersonic vehicle and another paper on digital twins. Both papers present innovative concepts in the objective of space exploration. As the technological ability of the world increases, there is evidence that many countries are increasing their research activities in space. The number of papers in this conference is encouraging.

The remaining 8 papers are grouped under mechanical engineering theme. Typical mechanical papers such as the lifting mechanism and tie rod of landing gears involve analysis of vibration or shock to mechanisms. It is also nice to see more fundamental mechanical engineering problems such as the research on flatness of large deployable antenna are investigated. Likewise, the research investigating flow-induced vibration of prisms has good potential to develop into more accurate prediction of mechanism interaction with fluid flow.

The research rigour and strong theoretical fundamentals demonstrated in the publications in this conference make this proceeding a valuable repository of research topics in aeronautical, aerospace and mechanical engineering. I would like to thank the conference committee, the conference secretariat, keynote speakers and all participants for their hard work making this conference a success.

Professor John P.T. Mo
Conference Chair, AAME 2020
BSc(Engg), MSc(Engg), PhD(Loughborough), FIEAust CPEng NER APEC Engineer IntPE(Aus), CEng, FIMechE, MHKIE, MFIE
School of Engineering
RMIT University

February 2020

IOP Conf. Series: Materials Science and Engineering **816** (2020) 011002 doi:10.1088/1757-899X/816/1/011002

Organizing Committees

Conference Chairs

Prof. Renfu Li, Huazhong University of Science and Technology, China
Prof. John Mo, Royal Melbourne Institute of Technology, Australia

Technical Program Chair

Prof. Petr Valášek, Czech University of Life Science Prague, Czech Republic

Technical Committees

Prof. Ping Yang, Jiangsu University, China
Prof. Hariati Taib, Universiti Tun Hussein Onn Malaysia, Malaysia
Prof. Qi Wang, Nanchang Hangkong University, China
Assoc. Prof. Eriki Ananda Kumar, FTVETI University, Ethiopia
Assoc. Prof. Baharudin B.T.H.T, Universiti Putra Malaysia, Malaysia
Assoc. Prof. Ts. Dr. Devarajan Ramasamy, Universiti Malaysia Pahang, Malaysia
Assoc. Prof. K. Kadirgama, Universiti Malaysia Pahang, Malaysia
Assoc. Prof. Dr. Yamuna Munusamy, Universiti Tunku Abdul Rahman, Malaysia
Assoc. Prof. Rusdi Nur, Politeknik Negeri Ujung Pandang, Indonesia
Asst. Prof. Min Zhou, Northwestern Polytechnical University, China
Asst. Prof. Nolan Concha, FEU-Institute of Technology, Philippines
Asst. Prof. Dr. Sudheer A P, National Institute of Technology Calicut, India
Dr. S. R. Jufar, Universiti Teknologi PETRONAS, Malaysia
Dr. Ramasamy, Universiti Malaysia Pahang, Malaysia
Dr. Muhamad Mat Noor, University Malaysia Pahang, Malaysia
Dr. Simon Winberg, University of Cape Town, South Africa

IOP Conf. Series: Materials Science and Engineering **816** (2020) 011003 doi:10.1088/1757-899X/816/1/011003

First Name	Last Name	Institution	Email Address
Xiaoyan	Liu	Shanghai Aircraft Design And Research Institute, COMAC,China	liuxiaoyan@comac.cc
Atik	Morshed	Nanchang Hangkong University, China	34005@nchu.edu.cn
Micheng	Zha	First author,China	354955897@qq.com
Guqi	Peng	Technology and Engineering Center for Space Utilization,China	pengguqi17@csu.ac.cn
Fufu	Wang	Key Laboratory of Space Utilization, Technology and Engineering Center for space Utilization, Chinese Academy of Sciences,China	wangff16@csu.ac.cn
Bo	Yang	Unit 96815 of PLA,China	yang_bo_nudt@163.com
Qian	Zhang	Air Force Engineering University,China	zhangqian8913@163.com
XIAO	YANG	Tsinghua University,China	yangxiao502@126.com
Ze	Zhang	Air Force Engineering University,China	zhangze_xyz@163.com
LaOdeAhmad	Barata	Kanazawa University,Japan	ahmad.barata@uho.ac.id
Yunke	Zhao	Beihang University,China	zhaoyunke@buaa.edu.cn
YUN	LI	China Academy of Space Technology,China	handanliyun@126.com
jinming	Chen	Nanchang Hangkong University, Nanchang,CHINA	1160475603@qq.com
Hang	Song	Nanchang Hangkong University,CHINA	sh_3212756@163.com
lei	Wang	Nanchang Hangkong University,CHINA	1635849403@qq.com
Mingtong	Tang	Nanchang Hangkong University,China	645172379@qq.com
Qidong	Li	Nanjing university of aeronautics and astronautics,China	1529320434@qq.com
Zeyu	Li	Beihang University,China	zeyulee@buaa.edu.cn
Shichao	Guo	Beijing Institute of Spacecraft Environment Engineering,China	357464199@qq.com
Yanni	Wei	Xi'an University of Technology,China	weiyanni@xaut.edu.cn
Longbin	LIU	College of Aerospace Science and Engineering, National University of Defense Technology,China	longbuaa@163.com
Ying	Li	Beijing Institute of Tracking and Telecommunications Technology,China	liying@bittt.cn
YANG	LIU	shanghai aircraft design and research institute, China	auvpomelo@126.com

Peer review statement

All papers published in this volume of *IOP Conference Series: Materials Science and Engineering* have been peer reviewed through processes administered by the proceedings Editors. Reviews were conducted by expert referees to the professional and scientific standards expected of a proceedings journal published by IOP Publishing.

IOP Conf. Series: Materials Science and Engineering **816** (2020) 012001 doi:10.1088/1757-899X/816/1/012001

Research on Vibration Suppression for AGARD Wing Based on State Space Model

Xiaoyan Liu

Shanghai Aircraft Design and Research Institute, COMAC, No.5188 Jinke Road, Pudong District, Shanghai, China

liuxiaoyan@comac.cc

Abstract. Based on the Observer method, the unsteady aerodynamic reduced order model is established. The aerodynamic, structural and control parts are in the form of state space equations. The state feedback control law is designed by linear quadratic regulator (LQR) method, and the active flutter suppression of aeroelastic system is realized. The three-dimensional AGARD elastic wing is selected as an example. The simulation results show that the active control law designed by the model reduction technique can effectively stabilize the response of the system, and the critical speed of closed-loop flutter increases greatly.

1. Introduction

The design of modern aeroelastic active control system requires a low-order state-space model. The computational fluid dynamics (CFD) method has considerable advantages in the simulation of nonlinear problems, but the engineering application is limited to a certain extent. It is not suitable for active control law design and conventional analysis. Therefore, it is necessary to apply the unsteady aerodynamic reduced order model (ROM) with high computational efficiency to the aeroelastic control system. The aerodynamic ROM reflects the important characteristics of the aerodynamic system, and its accuracy is equivalent to that of the CFD solver. It can be integrated into the conventional analysis of aeroelastic system control. Unsteady aerodynamic ROM can replace CFD solver, which is used for time domain simulation and integrated design of aeroelastic system.

The flutter critical velocity of many aircraft in transonic velocity region will appear a unique "pit" phenomenon, which directly restricts the flight envelope of the aircraft. Active flutter suppression is to transform unstable mode branches into stable ones by closed-loop control, and then improve flutter characteristics of aircraft and enlarge flight envelope. Usually, the aerodynamic control surface is used as the actuator which is deflected to generate additional unsteady aerodynamic force, and the energy of the aeroelastic system is balanced in the form of artificial damping. Reference [1] establish aerodynamic ROM by system identification method. In Matlab Simulink active flutter suppression of BACT wing with plunge/ pitching degrees of freedom is carried out. Reference [2] used the ARMA method to establish the active suppression model of the aero-servo-elastic(ASE) system by coupling structural state equations, transonic unsteady aerodynamic ROM and the aerodynamic model of the control surface. The suboptimal control method based on output feedback is used to design the control law. Reference[3] used Volterra series ROM to carry out ASE synthesis design for two-dimensional BACT wing. Compared with CFD/CSD direct simulation, the computational efficiency of the reduced-order model was improved by 2 ~ 3 orders of magnitude. Reference [4] used POD method and equilibrium cut-off method to obtain aerodynamic ROM and reduced-order aeroelastic system, and designed active control law based on LQR method to achieve flutter suppression for wing segments.

The low order model suitable for active control system design has certain requirements for the number of degrees of freedom, the cost of model generation, robustness and precision. In this paper, the Observer method [5] is used to establish the aerodynamic reduced order model. The physical smooth input is selected to train the CFD model so that it has a high computational efficiency in the time domain and the construction cost of the model is relatively small. The reduced order aeroelastic control system is constructed in the form of state space equations for aerodynamic, structural and control parts. The standard elastic model AGARD wing is taken as the object, and LQR method is used to design the active flutter suppression through the state space model.

2. Construction of Aerodynamic Rom

2.1. Overview of Method

System impulse response sampling is also called Markov parameter. Extracting Markov parameters from input / output data is an important step in system identification. These parameters can be used to develop efficient state space models of aeroelastic systems. The Observer method has been successfully used to identify modal parameters of structural systems. The Markov parameters of the small damping system decay slowly. First, the Markov parameters of the corresponding observer are obtained, and then the Markov parameters of the system itself are recovered [6]. The system is indirectly identified by asymptotically stable observer model and pole assignment method [7]. Ref. [8] discussed the pole assignment method of the observer, whose value can be real number, complex number or zero (corresponding to the Deadbeat observer).

In this paper, the Observer method is extended to the fluid system as in [5]. Discrete linear state space models are represented as

$$\begin{aligned} \boldsymbol{x}(k+1) &= \boldsymbol{A}\boldsymbol{x}(k)+\boldsymbol{B}\boldsymbol{u}(k) \\ \boldsymbol{y}(k) &= \boldsymbol{C}\boldsymbol{x}(k)+\boldsymbol{D}\boldsymbol{u}(k) \end{aligned} \tag{1}$$

When the above system is at zero initial condition $\boldsymbol{x}(0)=0$, its input/output relation is [8]

$$\boldsymbol{y}(k)=\sum_{\tau=0}^{k-1}\boldsymbol{Y}_{k-\tau-1}\boldsymbol{u}(\tau)+\boldsymbol{D}\boldsymbol{u}(k)\,. \tag{2}$$

$\boldsymbol{Y}_{k-\tau-1}$ is denoted by the product $\boldsymbol{C}\boldsymbol{A}^{k-\tau-1}\boldsymbol{B}$, Equation (1) is written in matrix form about the time series k [6]

$$\underset{q\times l}{\boldsymbol{y}} = \underset{q\times ml}{\boldsymbol{Y}}\ \underset{ml\times l}{\boldsymbol{U}} \tag{3}$$

$$\boldsymbol{Y}=\begin{bmatrix}\boldsymbol{D} & \boldsymbol{CB} & \boldsymbol{CAB} & \cdots & \boldsymbol{CA}^{l-2}\boldsymbol{B}\end{bmatrix}$$

$\boldsymbol{Y}$ includes all the Markov parameters that need to be solved. For small damping systems, the scale of $\boldsymbol{U}$ may be too large to obtain the pseudo inverse $\boldsymbol{U}^{+}$ by numerical calculation. The observer method solves this problem by introducing an observer into the input / output relation matrix, whose eigenvalue can be arbitrarily specified. Add and subtract $\boldsymbol{G}\boldsymbol{y}(k)$ at the right side of (1), then the observer model is constructed as [6]

$$\begin{aligned} \boldsymbol{x}(k+1) &= \bar{\boldsymbol{A}}\boldsymbol{x}(k)+\bar{\boldsymbol{B}}\boldsymbol{v}(k) \\ \boldsymbol{y}(k) &= \boldsymbol{C}\boldsymbol{x}(k)+\boldsymbol{D}\boldsymbol{u}(k) \\ \bar{\boldsymbol{A}} = \boldsymbol{A}+\boldsymbol{G}\boldsymbol{C}\,, &\quad \bar{\boldsymbol{B}}=\begin{bmatrix}\boldsymbol{B}+\boldsymbol{G}\boldsymbol{D}, & -\boldsymbol{G}\end{bmatrix} \\ \boldsymbol{v}(k) &= \begin{bmatrix}\boldsymbol{u}(k) \\ \boldsymbol{y}(k)\end{bmatrix} \end{aligned} \tag{4}$$

$\boldsymbol{G}$ is an arbitrary matrix of $n\times q$. If the original system is observable, the purpose of $\boldsymbol{G}$ is to configure the eigenvalue of matrix $\bar{\boldsymbol{A}}$ to be stabilized. At this point, the corresponding Markov parameter is called observer Markov parameter, and the input / output relationship is [6]

$$y(k)=\sum_{\tau=0}^{k-1}\bar{Y}_{k-\tau-1}v(\tau)+Du(k) \tag{5}$$

Where

$$\bar{Y}_{k-\tau-1}=C\bar{A}^{k-\tau-1}\bar{B} \tag{6}$$

The Markov parameters of the original system are restored based on the Deadbeat observer, and the unsteady aerodynamic state space model is constructed by using the Markov parameters combined with the eigenvalue realization algorithm.

2.2. Validation of Method

In order to verify the aerodynamic modeling method based on Observer technology, a AGARD CT6 example [9] is selected, which is often cited in many references. The airfoil is subjected to forced torsional motion around 1/4 chord length point. The CFD numerical simulation is carried out by using the N-S equations along with the dynamic grid technique, with $M_\infty=0.796$ is simulated. The variation of lift coefficient with attack of attack is shown in figure 1.

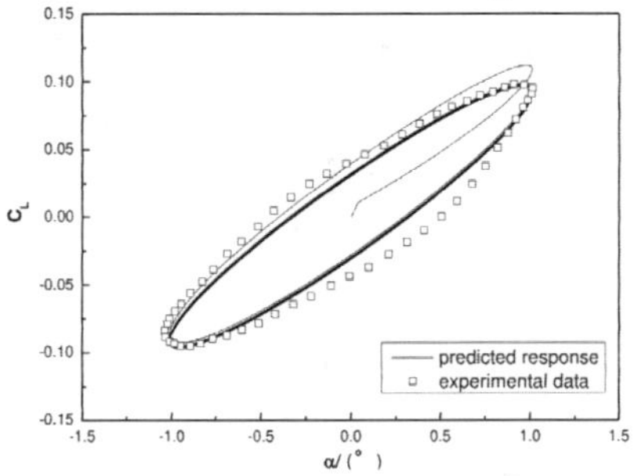

Figure 1. Comparison of lift coefficient between ROM and experiment.

The reduced-order model shows the nonlinear effect of aerodynamic force reasonably. The predicted response is consistent with the wind tunnel experimental data [9], and retains the mainstream nonlinear characteristics. It only takes a few minutes to predict the response of the unsteady aerodynamics by means of state space model.

3. Design of Aeroelastic Active Control Law

3.1. Aeroelastic Control System Model

Coupling the aerodynamic state equations with structural state equations, aeroelastic system is written as

$$\begin{aligned} x_{open}(k+1)&=A_{open}x_{open}(k)+B_{open}F(k)\\ y_{open}(k)&=C_{open}x_{open}(k) \end{aligned} \tag{7}$$

The feedback control is realized by the active deflection of the control surface to generate additional aerodynamic forces. In this paper, it is assumed that the elastic deformation of the control surface relative to the main wing surface is negligible and the rigid body rotation of the control surface is only considered. In addition, the control surface is relatively smaller than the main wing surface, thus the interaction of aerodynamic interaction is not obvious. As a result, the unsteady aerodynamic force can be decomposed into both the main wing and the control surface[10]. An aerodynamic state equation corresponding to the mode of the control surface is established by means of the aerodynamic ROM technique

$$\begin{aligned} \boldsymbol{x}_c(k+1) &= \boldsymbol{A}_c\boldsymbol{x}_c(k)+\boldsymbol{B}_c\delta(k) \\ \boldsymbol{f}_c(k) &= \boldsymbol{C}_c\boldsymbol{x}_c(k)+\boldsymbol{D}_c\delta(k) \end{aligned} \tag{8}$$

Where δ is the deflection angle of the control surface, $\boldsymbol{f}_c$ is the additional aerodynamic coefficient caused by the deflection of the control surface.

Set $\boldsymbol{x}_{ase}=\left[\boldsymbol{x}_{open},\boldsymbol{x}_c\right]^{\mathrm{T}}$, $\boldsymbol{F}_c=q\boldsymbol{f}_c$, The open loop state equation of aeroelastic control system is obtained by joining equation (7) with equation (8), get

$$\begin{aligned} \boldsymbol{x}_{ase}(k+1) &= \boldsymbol{A}_{ase}\boldsymbol{x}_{ase}(k)+\boldsymbol{B}_{ase}\delta(k) \\ \boldsymbol{y}_{ase}(k) &= \boldsymbol{C}_{ase}\boldsymbol{x}_{ase}(k) \end{aligned} \tag{9}$$

Where

$$\boldsymbol{A}_{ase}=\begin{bmatrix} \boldsymbol{A}_{open} & q\boldsymbol{B}_{open}\boldsymbol{C}_c \\ \boldsymbol{0} & \boldsymbol{A}_c \end{bmatrix}, \quad \boldsymbol{B}_{ase}=\begin{bmatrix} q\boldsymbol{B}_{open}\boldsymbol{D}_c \\ \boldsymbol{B}_c \end{bmatrix}, \quad \boldsymbol{C}_{ase}=\left[\boldsymbol{C}_{open},\boldsymbol{0}\right].$$

3.2. Vibration suppression design

The AGARD 445.6 wing is a classical aeroelastic test example [11]. In this paper, the geometric shape of the model is selected and the structural data of the model are used. The trailing edge control surface is installed at the outboard wing, and the structure layout is shown in figure 2. The span of the control surface is 30% of the wing span and the chord length of the control surface is 25% of the wing chord.

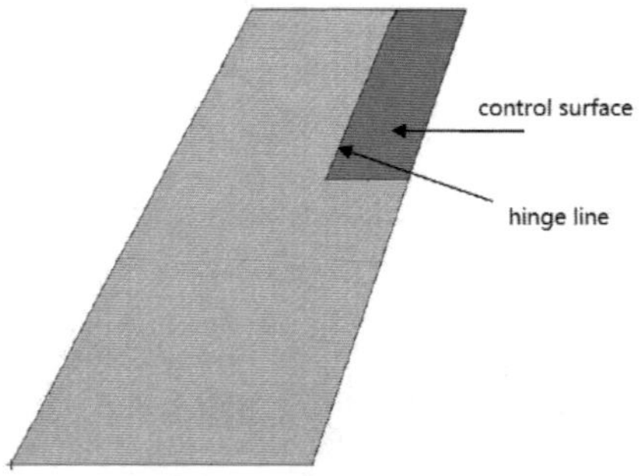

Figure 2. AGARD wing with control surface.

Firstly, the open-loop aeroelastic system is studied. The 70-order unsteady aerodynamic state equations of AGARD wing is constructed by using the Observer method, and the calculated open-loop flutter velocity is $V_f^*=0.4264$ when $M_\infty=0.678$. For $V^*=0.45$, the generalized coordinate response of the AGARD wing is shown in Fig 3, and the system is at an unstable divergent state. At this speed, the LQR method is used to design the feedback control, and the system flow chart is shown in Fig. 4. The Observer method is used to establish the 45-order aerodynamic model of the control surface. Coupling aerodynamic equation, structural equation and control surface equation, an open-loop aeroelastic control model is established which is a controlled object for design of control law.

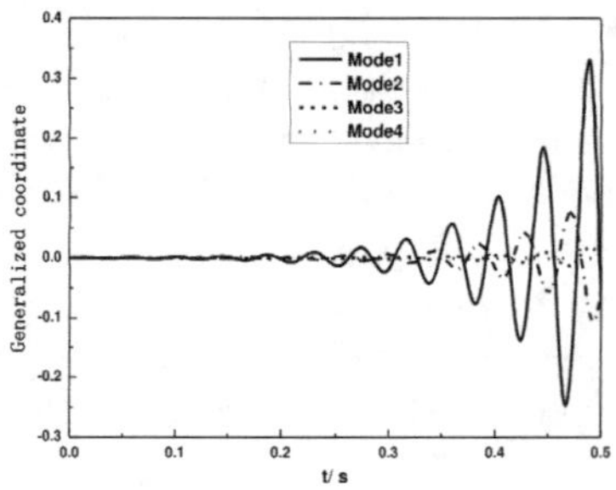

Figure 3. Generalized coordinate response of each mode without control.

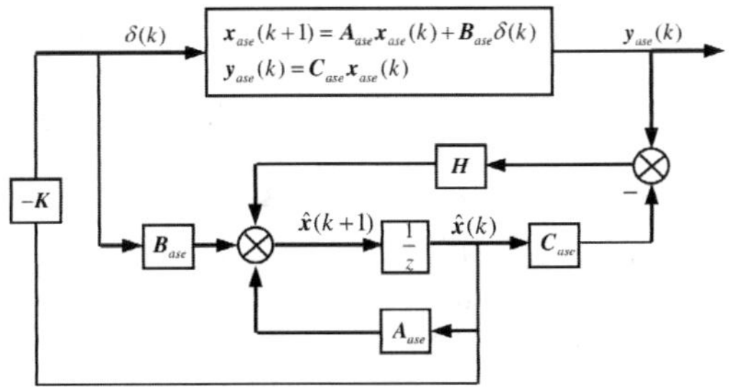

Figure 4. system flow chart based on LQR method.

In closed loop case, the modal response of wing structure are shown in figure 5. The response decays rapidly after initial disturbance, which indicates that the stability of the system is improved obviously under the control system. Change V^* until the aeroelastic control system reaches the critical state. The closed-loop critical response is shown in figure 6 and figure 7 The critical $V_f^* = 0.5043$ The flutter velocity is 20.8% higher than the uncontrolled wind tunnel test value 0.4174[11].

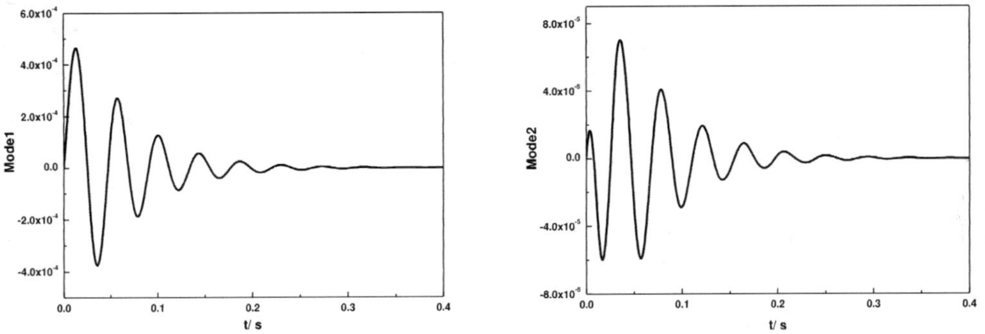

a) MODE1 controlled response. b) MODE2 controlled response.

Figure 5. system modal responses.

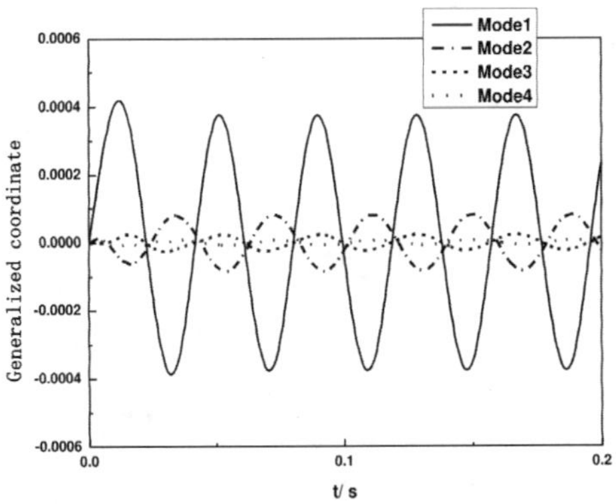

Figure 6. Critical responses of generalized coordinates of closed-loop modes.

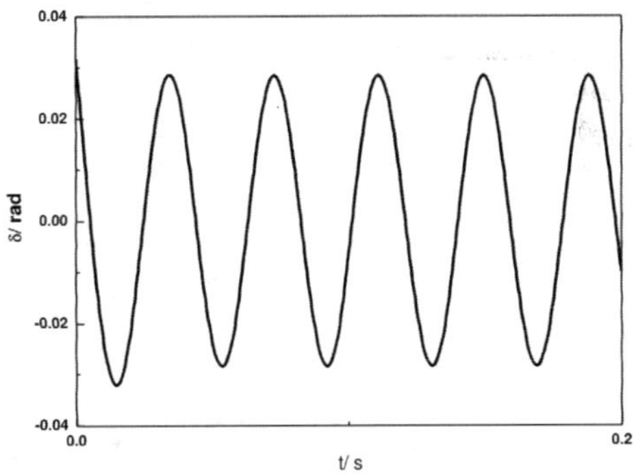

Figure 7. Response of control surface in closed-loop critical state.

3.3. Aeroelastic control system reduction by equilibrium truncation

The order of the three-dimensional aeroelastic control model is relatively high, and it is still necessary to reduce the order of the model in the engineering application of the control law design. In this paper, the balanced truncation method [12] is adopted. The controllable and observable Gramian matrix of the system is defined as [12]

$$\mathbf{W}_c = \int_0^\infty e^{At}\mathbf{B}\mathbf{B}^{\mathrm{T}}e^{A^{\mathrm{T}}t}dt$$
$$\mathbf{W}_o = \int_0^\infty e^{A^{\mathrm{T}}t}\mathbf{C}^{\mathrm{T}}\mathbf{C}e^{At}dt \tag{10}$$

It is generally considered that the actual system is controllable and stable, then $\mathbf{W}_c$ and $\mathbf{W}_o$ are nonsingular, symmetric positive semidefinite matrix. The realization of the internal balance of the system is to find the nonsingular transformation matrix $\mathbf{T}$ by which the original system $\{\mathbf{A},\mathbf{B},\mathbf{C},\mathbf{D}\}$ is transformed to $\{\mathbf{A}_t,\mathbf{B}_t,\mathbf{C}_t,\mathbf{D}\}$, and the controllable and observable Gramian matrix is transformed into the same diagonal matrix

$$\mathbf{W}_{cb} = \mathbf{W}_{ob} = \Sigma = \mathrm{diag}(\sigma_1^2 \quad \sigma_2^2 \quad \cdots \quad \sigma_n^2)\,. \tag{11}$$

Where σ_i is the Hankel singular value that reflects the comprehensive controllability and observability of the system state, $\sigma_1 \ge \sigma_2 \ge \cdots \ge \sigma_n$ .Truncate the state variable with lower controllability and observability, then the original system becomes $\{\mathbf{A}_{tm},\mathbf{B}_{tm},\mathbf{C}_{tm},\mathbf{D}\}$.

The equilibrium truncation reduction of the open-loop aeroelastic model at Mach number 0.678 is presented in this paper. The full-order model is composed of 200-order aerodynamic state equation, 90-order control surface aerodynamic equation and 8-order structural state equation. The full-order model is reduced to 80-order, 50-order, 40-order and 30-order respectively. The open-loop flutter velocity calculated by full-order model is 0.4195 and the corresponding wind tunnel test value is 0.4174. The comparison of open-loop flutter response between full-order model and 30-order truncated model are given in Fig.8. Design of flutter suppression control law based on truncated reduced order aeroelastic control model, and it would be applied to full-order model. The results are shown in Table 1. From the results of flutter suppression, the closed-loop flutter velocity of different order truncated model is close, but the increase ratio is obviously lower than that of direct design of full-order model.

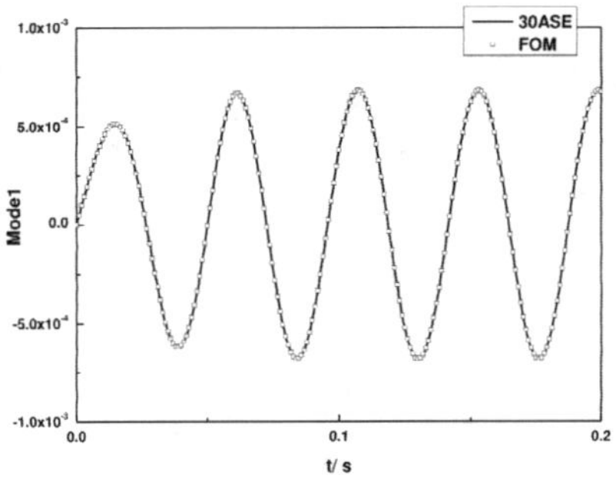

a) MODE1 response.

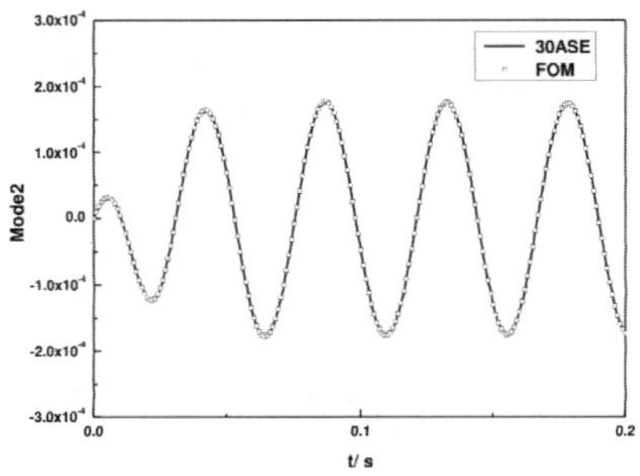

b) MODE2 response.

Figure 8. Comparison of open-loop flutter responses.

Table 1. Closed -loop results

System order	Closed-loop flutter velocity	Higher than calculated open loop velocity
Full order 298	0.5236	24.8%
80	0.4501	7.3%
50	0.4526	7.9%
40	0.4501	7.3%
30	0.4519	7.7%

4. Conclusions

The Observer method focuses on the external input / output relationship of the system, extracts the Markov parameters of the system, and directly establishes the state space equation of the system. This form is beneficial to the comprehensive design of the aeroelastic control system. In this paper, the Observer method is introduced into identification of transonic unsteady aerodynamics, and the reduced-order aerodynamic model with higher accuracy and lower order is established, which is coupled with structural and control equations to construct the state-space model of low-order aeroelastic control system that is convenient for engineering application. It provides a theoretical basis for the engineering realization of transonic low-order flutter active suppression. The transonic AGARD wing is selected as an example. The simulation results show that the low-order aeroelastic control model is suitable for active flutter suppression in engineering field.

5. References

[1] Stephens C. H., Arena S. A. J., Gupta K. K. CFD-Based Aeroservoelastic Predictions with Comparisons Benchmark Experimental data. *AIAA 99-0766*, 1999
[2] Zhang Weiwei, Ye Zhengyin. Active suppression of transonic flutter. *J of Vibration Engineering*, 2007,**20** (3): 224-231
[3] Yao Weigang, Xu Min. Aeroelastic analysis based on reduced order model of Volterra series. *J of Aeronautics and Astronautics*, 2008, **29** (6): 1711 -1716
[4] Xu Min, Anxiaomin, Zeng Xianang, et al. Aeroelastic reduction based on BPOD and its application in active control. *Chinese sci-tech papers online*, 2008, **3** (10): 781-786
[5] Liu Xiaoyan, Wu Zhigang, Yang Chao. Aerodynamic Reduced–Order Models Based on Observer Techniques. *AIAA Paper 2010-2794*, 2010.
[6] Juang J-n, Phan M., Horta L. G., et al. Identification of Observer/Kalman Filter Markov Parameters: Theory and Experiments. *J of Guidance, Control, and Dynamics*, 1993, **16**(2): 320-329
[7] Phan M., Horta L. G., Juang J-n, et al. Identification of Linear Systems by an Asymptotically Stable Observer. *NASA Tech*. Paper 3164, 1992
[8] Phan M., Horta L. G., Juang J-n, et al. Linear System Identification via an Asymptotically Stable Observer. *J of Optimization Theory and Application*, 1993, **79**(1): 59-86
[9] Davis S. S. "NACA64a010 Oscillatory Pitching", in Compendium of Unsteady Aerodynamics Measurements. *AGARD-R*-702, 1982
[10] Xu Min, Chen Gang, Chen Shilu, etc. Design of Aeroelastic active Control Law based on unsteady aerodynamic low-order Model. *J of Northwestern University of Technology*, 2004, **22** (6): 748 -752
[11] Yates E. C. Jr. AGARD Standard Aeroelastic Configurations for Dynamic Response Candidate Configuration I.-Wing 445.6. *NASA-TM*-100492, 1987
[12] Moore B C. Principal Component Analysis in Linear System: Controllability, Observability and Model Reduction. *IEEE Trans, Automatic Control,* 1981, AC-26(1): 17-32

Fibrous Orientation Effects on the Deformation and Stress of a Pigeon -like Flapping-wing Micro Aerial Vehicle Based on APDL

A K M Atik Morshed, Chunjin Yu*, Fang Wang

Institute of Aircraft Engineering, Nanchang Hangkong University, Nanchang 330063, China

E-mail: 34005@nchu.edu.cn

Abstract: Fibrous orientation effects on the deformation and stress of a bird-like (pigeon) composite wing was studied. The bird wing is simulated and modeled by mechanical APDL software. The stress and displacement results of bird wing model under different laying modes are obtained by finite element method. The optimal laying mode is analyzed. The results are checked by strength theory in composite mechanics. The results show that for singe orientation $[-20]_4$ is the best orientation. For 4 layers structure, [-45/90/0/45] has best performance in 1st layer while [45/0/-45/90] has worst performance; [0/-45/90/45] has best performance in 2nd layer while [0/45/-45/90] has worst performance. [45/90/-45/0] has best performance for pigeon wing in this design in 3rd layer while [-45/0/90/45] has worst performance. [45/0/90/-45] has best performance for pigeon wing in this design in 4th layer while [-45/90/0/45] has worst performance. By comparison, it is found that the different-direction pavement is better than the same-direction pavement.

Introduction:

Wright Brothers watched birds flying basics for 10 years to figure out the basics of flying and finally for the first time became successful to build the first flying machine. After that throughout the century flying machines have been developed a lot and even developing faster. But this paper is about very different flying machine's wing. When researchers are talking about flying machine its wing is the most important part. Actually, from the beginning of flying idea people started to think built flying machine with flapping wings. Although, the flapping wing isn't very effective for the larger aircrafts but it has open new era for Micro Air Vehicles (MAV). It will take the generation of drone in different level. Compare to flapping wings, conventional fixed wing airplane are relatively simple [1-4].

Past few years' scientist have seen the development of flapping wings in Micro Air Vehicles, their research is inspired by small insects and birds' wings. Due to its small size and shape engineers have to use composite materials. The flapping wing is promising for its consistency, ability and elasticity. It can fly the MAV in very difficult situation and it can fly or hover to places where other aircrafts can't fly. These MAV with flapping wings can almost fly under any circumstance. The manufacturing cost is also very low. These FWMAV could be use in different sector in human civilization. It could be used in civil, military and research purpose. For example, it could be used to take a picture where human can't go or use in a rescue mission or espionage or surveillance to gather information. It promising because of its ability of flying in critical situation and ability of carrying load compare to its size. Although the recent knowledge about flapping wings is not too much. Also, the dynamic of these kind of wing isn't fully understood. But it's believed that when a composite material flapping wing is scaled down to a small bird wing size it is more efficient. Compared to conventional types of aircraft, such as rotorcraft and fixed-wing vehicles, FWMAVs are believed more efficient when scaled down to the insect size range [5]. Wing membranes are highly flexible with large deformations during the stroke [6].

A major potential advantage claimed for fibrous composite materials in structural applications is that the material can be tailored by proper orientation of the fibers in various layers so as to optimize the desired structural behavior [7]. And yet fibrous composites nature is still unpredictable which kept the optimal tailoring beyond the most design engineers.

FWMAVs is an imitation of birds or insects [8-10] flying new aircraft concepts in the application of its technology beyond the traditional scope of the study aircraft design and aerodynamics. [11-13], while creating micro-electromechanical systems (MEMS) technology applications in the field of aviation. Bionic efficient wing design and manufacture with good dynamic characteristics [14-15], is bionic FMAV study of a key, surely it has a very bright future and it can do a lot in development of drone technology but also the currently very challenging research problems. In this paper fibrous orientation effects of a bird-like (pigeon) composite wing will be studied.

Method of Calculation

Stress-strain relationship between simple laminates in any direction can be represented by stresses in a 1-2 coordinate system, and the conversion equation is:

$$\begin{Bmatrix}\sigma_x\\ \sigma_y\\ \tau_{xy}\end{Bmatrix}=\begin{bmatrix}cos^2\theta & sin^2\theta & -2sin\theta cos\theta\\ sin^2\theta & cos^2\theta & 2sin\theta cos\theta\\ sin\theta cos\theta & -sin\theta cos\theta & cos^2\theta-sin^2\theta\end{bmatrix}\begin{Bmatrix}\sigma_1\\ \sigma_2\\ \tau_{12}\end{Bmatrix} \tag{1}$$

$$\begin{Bmatrix}\sigma_1\\ \sigma_2\\ \tau_{12}\end{Bmatrix}=\begin{bmatrix}cos^2\theta & sin^2\theta & 2sin\theta cos\theta\\ sin^2\theta & cos^2\theta & -2sin\theta cos\theta\\ -sin\theta cos\theta & sin\theta cos\theta & cos^2\theta-sin^2\theta\end{bmatrix}\begin{Bmatrix}\sigma_x\\ \sigma_y\\ \tau_{xy}\end{Bmatrix} \tag{2}$$

Hence, m=cosθ,n=sinθ, Then,

$$\begin{Bmatrix}\sigma_1\\ \sigma_2\\ \tau_{12}\end{Bmatrix}=\begin{bmatrix}m^2 & n^2 & 2mn\\ n^2 & m^2 & -2mn\\ -mn & mn & m^2-n^2\end{bmatrix}\begin{Bmatrix}\sigma_x\\ \sigma_y\\ \tau_{xy}\end{Bmatrix} \tag{3}$$

We can write as, $\{\sigma_1\}=[T_\sigma]\{\sigma_X\}$, **Here, the stress conversion matrix is called the $[T_\sigma]$.**

Strain conversion formula

$$\begin{Bmatrix}\varepsilon_x\\ \varepsilon_y\\ \gamma_{xy}\end{Bmatrix}=\begin{bmatrix}cos^2\theta & sin^2\theta & -2sin\theta cos\theta\\ sin^2\theta & cos^2\theta & 2sin\theta cos\theta\\ sin\theta cos\theta & -sin\theta cos\theta & cos^2\theta-sin^2\theta\end{bmatrix}\begin{Bmatrix}\varepsilon_1\\ \varepsilon_2\\ \gamma_{12}\end{Bmatrix} \tag{4}$$

$$\begin{Bmatrix}\varepsilon_1\\ \varepsilon_2\\ \gamma_{12}\end{Bmatrix}=\begin{bmatrix}cos^2\theta & sin^2\theta & 2sin\theta cos\theta\\ sin^2\theta & cos^2\theta & -2sin\theta cos\theta\\ -sin\theta cos\theta & sin\theta cos\theta & cos^2\theta-sin^2\theta\end{bmatrix}\begin{Bmatrix}\varepsilon_x\\ \varepsilon_y\\ \gamma_{xy}\end{Bmatrix} \tag{5}$$

Hence, m=cosθ,n=sinθ, Then,

$$\begin{Bmatrix}\varepsilon_1\\ \varepsilon_2\\ \gamma_{12}\end{Bmatrix}=\begin{bmatrix}m^2 & n^2 & 2mn\\ n^2 & m^2 & -2mn\\ -mn & mn & m^2-n^2\end{bmatrix}\begin{Bmatrix}\varepsilon_x\\ \varepsilon_y\\ \gamma_{xy}\end{Bmatrix} \tag{6}$$

We can write as, $\{\varepsilon_1\}=[T_\varepsilon]\{\varepsilon_X\}$, Here This is called a stress conversion matrix $[T_\varepsilon]$.

Results and Discussion

APDL is ANSYS Parametric Design Language, and it is used to analyze the composites structure. A composite laminate plate is made up of a 4-layers unit plate, the unit plate performance parameters as

IOP Conf. Series: Materials Science and Engineering **816** (2020) 012002 doi:10.1088/1757-899X/816/1/012002

shown in Table 1, Assume that, the mass of pigeon 1kg and the area of its wing area is 0.05m^2 so the pressure MAV is 196N/m^2.

Table 1 Unit board performance parameters

EX/MPa	EY/MPa	EZ/MPa	GXY/MPa	GYZ/MPa	GXZ/MPa	PRXY	PRYZ	PRXZ
1250	300	300	50	20	50	0.25	0.25	0.01

The total deformation of composite wing is shown in Fig.1, and from fig.1 we know that the deformation increases with the increasing of spanwise length.

Fig.2 is the total deformation graph, orientation of composites is from $[-80]_4$ to $[80]_4$ degrees, and from fig.2 we know that when the orientation is [-20]4 has the best performance.

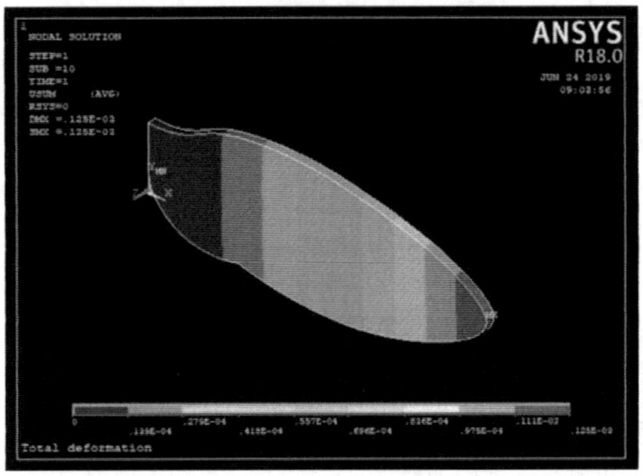

Figure 1. Total deformation of composite wing

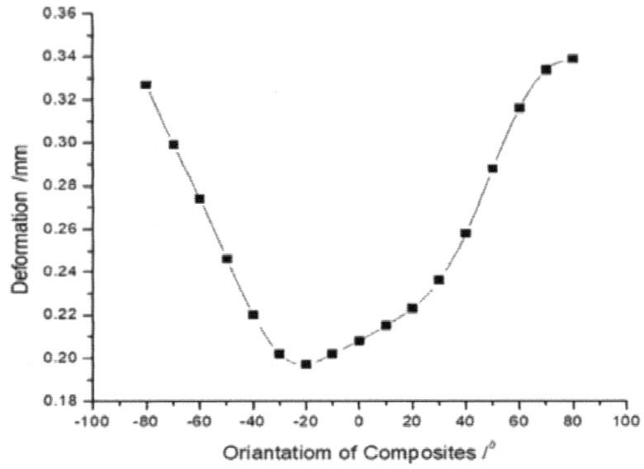

Figure.2. Deformation graph degrees $[-80]_4$ to $[80]_4$

Fig.3 is the total deformation graph, orientation of 24 cycles from [0/45/-45/90] to [-45/0/45/90];

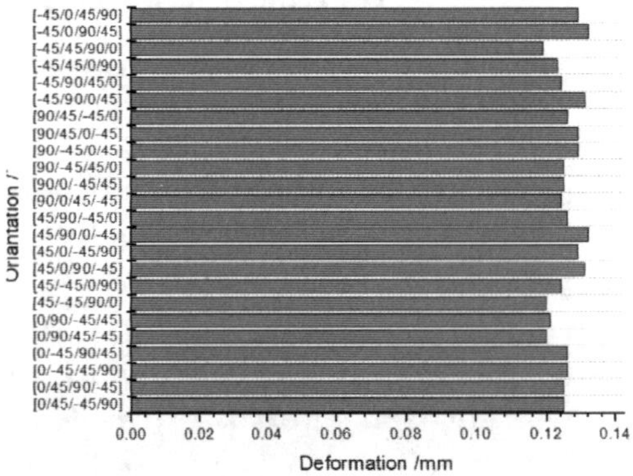

Figure.3. Deformation graph analysis of 24 orientation

Here, [0/90/45/-45] angle is best performance for pigeon wing in this design. Because [0/90/45/-45] angle is lowest deformation. In addition, [-45/0/90/45] angle is worst performance because [-45/0/90/45] angle have highest deformation.

Fig.4 is the stress graph of 1st layer. Orientation of 24 cycles from [0/45/-45/90] to [-45/0/45/90]. Here, [-45/90/0/45] orientation is best performance for pigeon wing in this design in 1st layer. Because [-45/90/0/45] is lowest stress. In addition, [45/0/-45/90] angle is worst performance because [45/0/-45/90] angle have highest stress in 1st layer.

Fig.5 is the stress graph of 2nd layer. Orientation of 24 cycles from [0/45/-45/90] to [-45/0/45/90]. Here, [0/-45/90/45] angle is best performance for pigeon wing in this design in 2nd layer. Because [0/-45/90/45] angle is lowest stress. In addition, [0/45/-45/90] angle is worst performance because [0/45/-45/90] angle have highest stress in 2nd layer.
Fig.6 is the stress graph of 3rd layer. Orientation of 24 cycles from [0/45/-45/90] to [-45/0/45/90]. Here, [45/90/-45/0] angle is best performance for pigeon wing in this design in 2nd layer. Because [45/90/-45/0] angle is lowest stress. In addition, [-45/0/90/45] angle is worst performance because [-45/0/90/45] angle have highest stress in 3rd layer.
Fig.7 is the stress graph of 4th layer. Orientation of 24 cycles from [0/45/-45/90] to [-45/0/45/90]. Here, [45/0/90/-45] angle is best performance for pigeon wing in this design in 4th layer. Because [45/0/90/-45] angle is lowest stress. In addition, [-45/90/0/45] angle is worst performance because [-45/90/0/45] angle have highest stress in 4th layer.

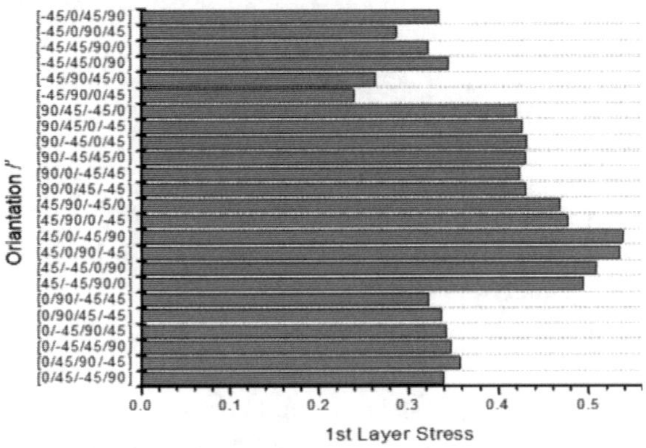

Figure.4. Stress graph analysis of 24 orientation in 1st layer

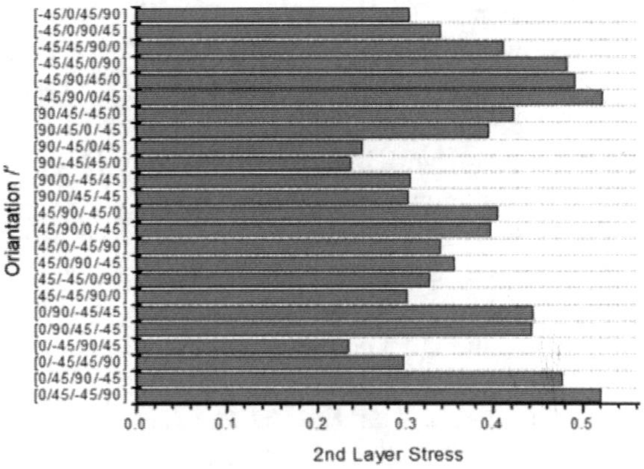

Figure.5. Stress graph analysis of 24 orientation in 2nd layer

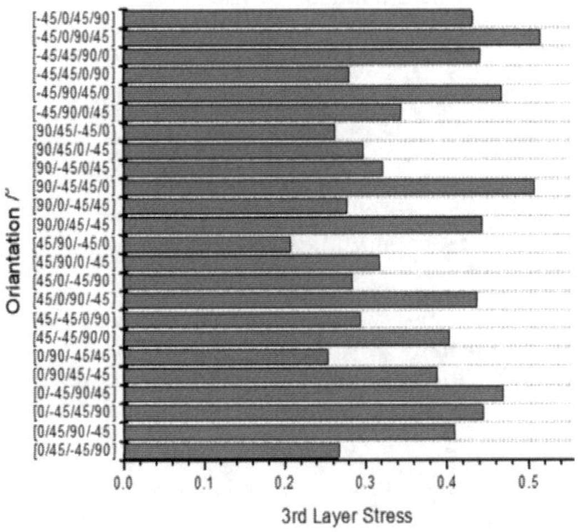

Figure.6. Stress graph analysis of 24 orientation in 3rd layer

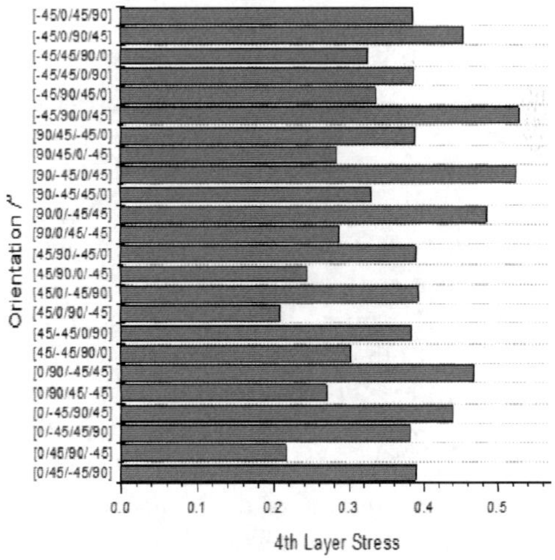

Figure.7. Stress graph analysis of 24 orientation in 4th layer

Conclusions

The representative of pigeon-like wing was studied as the research object. First a precise geometric model of pigeon wing was been established. Reference to the previous research results, the selection of

material was taken named as epoxy carbon. This material using for four layers epoxy carbon used for top and bottom layers and resin epoxy used in the middle layer resin epoxy is glue it is for adhesive top and bottom layers.
This paper analyzed total maximum deformation, maximum stress for four layers for single orientation angle. In addition, for singe orientation $[-20]_4$ is the best orientation. For 4 layers structure, [-45/90/0/45] has best performance in 1st layer while [45/0/-45/90] has worst performance; [0/-45/90/45] has best performance in 2nd layer while [0/45/-45/90] has worst performance. [45/90/-45/0] has best performance for pigeon wing in this design in 3rd layer while [-45/0/90/45] has worst performance. [45/0/90/-45] has best performance for pigeon wing in this design in 4th layer while [-45/90/0/45] has worst performance.

Reference

[1] Nguyen A T, Han J. Wing flexibility effects on the flight performance of an insect-like flapping wing micro-air vehicle. Aerospace science and Technology, 2018, 79: 468-481.
[2] Croon G C H E D, Clercq K M E D, Ruijsink R, et al. Design, aerodynamics, and vision-based control of the DelFly[J]. International Journal of Micro Air Vehicles, 2009, 1(2):71-97.
[3] Keennon M, Klingebiel K, Won H. Development of the Nano Hummingbird: A Tailless Flapping Wing Micro Air Vehicle[C]// Aiaa Aerospace Sciences Meeting Including the New Horizons Forum & Aerospace Exposition. 2012.
[4] Ma K Y, Chirarattananon P, Fuller S B, et al. Controlled Flight of a Biologically Inspired, Insect-Scale Robot[J]. Science, 2013, 340(6132):603-607.
[5] Floreano D, Wood R J. Science, technology and the future of small autonomous drones[J]. Nature, 2015, 521(7553):460-466.
[6] Shahzad A, Tian F B, Young J, et al. Effects of flexibility on the hovering performance of flapping wings with different shapes and aspect ratios[J]. Journal of Fluids and Structures, 2018, 81:69-96.
[7] Boukhoulda B F, Adda-Bedia E, Madani K. The effect of fiber orientation angle in composite materials on moisture absorption and material degradation after hygrothermal ageing[J]. Composite Structures, 2006, 74(4):406-418.
[8] KESEL A B, PHIL IPPI U, NACHTIGALL W. Biomechanical aspect s of the insect wing: an analysis using the finite element method[J] . Computers in Biology and Medicine, 2003 ,28 :4232437.
[9] Abas M F B, Rafie A S B M, Yusoff H B, et al. Flapping wing micro-aerial-vehicle: kinematics, membranes, and flapping mechanisms of ornithopter and insect flight[J]. Chinese Journal of Aeronautics, 2016, 29(5): 1159-1177.
[10] Chunlin G, Jiakun H, Zongjing Y, et al. Numerical investigation of the effects of different parameters on the thrust performance of three dimensional flapping wings[J]. Aerospace Science and Technology, 2019, 84: 431-445.
[11] Gong X, Bansmer S, Strobach C, et al. Deformation Measurement of a Birdlike Airfoil with Optical Flow and Numerical Simulation[J]. AIAA Journal, 2014, 52(12):2807-2816.
[12] Guo S, Li H, Zhou C, et al. Analysis and experiment of a bio-inspired flyable micro flapping wing rotor. Aerospace Science and Technology, 2018, 79: 506-517
[13] Liu L, Li H, Ang H, et al. Numerical investigation of flexible flapping wings using computational fluid dynamics/computational structural dynamics method[J]. Proceedings of the Institution of Mechanical Engineers Part G Journal of Aerospace Engineering, 2016, 232(1): 85-95
[14] Yu, C, Kim D and Zhao Y. Stress Analysis of Membrane Flapping-Wing Arial Vehicle Based on Different Material Models. Journal of Applied Mathematics and Physics, 2014, 2: 1023-1030.
[15] Mark J, Ziwen G, Shen I Y. The effect of structural deformation on flapping wing energetics[J]. Journal of Sound and Vibration, 2018, 429:176-192.

IOP Conf. Series: Materials Science and Engineering **816** (2020) 012003 doi:10.1088/1757-899X/816/1/012003

Numerical Study on the Extreme Impact Load of Wavy-Water Ditching

ZHAO Yunke[1], QU Qiulin[1], LIU Peiqing[1,*]

School of Aeronautic Science and Engineering, Beihang University, Beijing 100191, China

lpq@buaa.edu.cn

Abstract. In this paper, numerical method is used study the extreme impact load and its physical causes during the wavy aircraft ditching. the finite volume method and VOF method are used to capture the free surface, six degrees of freedom model and global motion mesh to deal with the relative motion between water and aircraft. and simulate the forced landing process of the aircraft. By choosing the place with the largest vertical velocity relative to the water as the impact position, the extreme impact load on the fuselage is predicted. In this paper, during the impact stage of the wavy water surface, the fuselage encountered an impact peak which is unforeseen on the calm water surface, and the magnitude of the impact peak was related to the sinking speed of the aircraft relative to the water surface. This paper also compares the movement attitude and overload history of aircraft ditching under five sea conditions with different wave height, and gives the influence rules of wave height on the extreme impact load and the peak values of other parameters, which provides a reference for the design of aircraft load distribution.

1. Introduction

With the increasing number of cross-water flight routes, the possibility that an aircraft encounters an accident in the air and needs to perform an emergency landing on water will also increase. The U.S. Department of Transportation stipulates that a planned water emergency landing is called ditching [1]. By executing a scheduled water emergency landing procedure, the longitudinal and lateral loads on the aircraft will be within the design range[2][3], and passengers will have several minutes of posture preparation time to withstand the impact.

At present, the research on aircraft ditching mainly focuses on calm water, which forms the general understanding of ditching. The process of ditching can be divided into four stages: contact, impact, planing and floating. The dramatic changes of extreme impact load and attitude during the ditching mainly occur in the stage of contact-impact, which usually ends in a few seconds and the entire process changes dramatically non-linearly. The common research methods for ditching include theoretical analysis, model experiment and numerical simulation.

The ditching belongs to the problem of water impact. The pioneer research work was achieved by Von Karman in 1929 when completing the analysis on the problem of water impact of seaplane [4], in which Von Karman used theoretical analysis method to analyse of body load during the impact. In 1932, Wagner improved the Von Karman's theory by using added mass theory [5], taking the influence of sur-face deformation of water in the process of impact into account. The study of characteristics of different shapes of the fuselage in the ditching is accomplished by McBride, Fisher et al. in 1953 [6], whose report contains 9 kinds of different shapes of the fuselage which have different effect on the water impact. After the 1980s, the numerical simulation method gradually replaced the model experiment method. The boundary element method [7] is applied to the vertical water impact of the simple object, and there is little related literature for complex three-dimensional objects impacting with larger speed. SPH method is used in the water entry problem calculation of the rigid body and elastic body. And due to the existence of relatively large horizontal velocity in the process of ditching which produces the positive pressure, negative pressure also appears because of the curvature of the aircraft shape. But at present, the SPH method that makes the constitutive equation as the control equation, which is relatively mature, can't simulate the negative pressure, so the SPH method that makes NS equation as the control equation should be developed if you want to apply SPH to the ditching problem. Serious mesh distortion and higher calculation cost will appear when using finite element method (FEM) to calculate the ditching problem [8]. Streckwall et al. [9] using VOF method to simulate the water-air interface to study the water impact characteristics of a single fuselage; Qu et al. [10] used the Global Motion Mesh (GMM) technology to process the relative motion between the aircraft and the water surface, and the numerical simulation results are in good agreement with the experimental results.

In the actual ditching process, the wave environment is often encountered. If an aircraft touches water in a dangerous wave position, there will be an unforeseen extreme impact load on a calm water surface. This paper uses numerical simulation method to study the ditching characteristics of a high-wing regional aircraft in wave environment with different wave height.

2. Numerical Method and Verification

2.1. Numerical methods

The computational approaches include four parts: flow solver, VOF method, six-degree-of-freedom (6DOF) model, and GMM method. The VOF method is used to capture free water surface. The 6DOF model is used to compute the motion trajectory of the object. The GMM method is used to deal with the motion of the whole computational domain.

In this paper, the whole computational domain (including the cells and boundary) moves together with the object like a rigid body. Thus there is no need to use any re-meshing or mesh deformation techniques, which will ensure the quality of the cells in the whole computational domain to improve numerical accuracy (especially the accuracy of free water surface capture) and save computation cost.

The volume fraction boundary conditions can ensure that the free water surface keeps a stationary level when the computational domain moves. This condition is set according to the cell coordinates of the boundaries in the earth fixed coordinate system. That is, the volume fraction of water $a = 0.5$ for the cell located on the interface between air and water; $a = 0$ for the cells located above the interface; $a = 1$ for the cells located below the interface.

2.2. Validation of numerical methods

The ditching process of NACA TN 2929 model F is simulated and compared with the experimental results [6]. In this case, a half mesh of Model F is used. Except the symmetry boundary and no-slip wall boundary, the other outer boundaries are set as velocity-inlet conditions. A mix of unstructured and Multi-block structure mesh of the total cell number of 2.7 million is used. The numerical results and the experimental results from reference are compared in figure 1. The numerical results of pitch attitude and CG height show a satisfactory agreement with the experiment results.

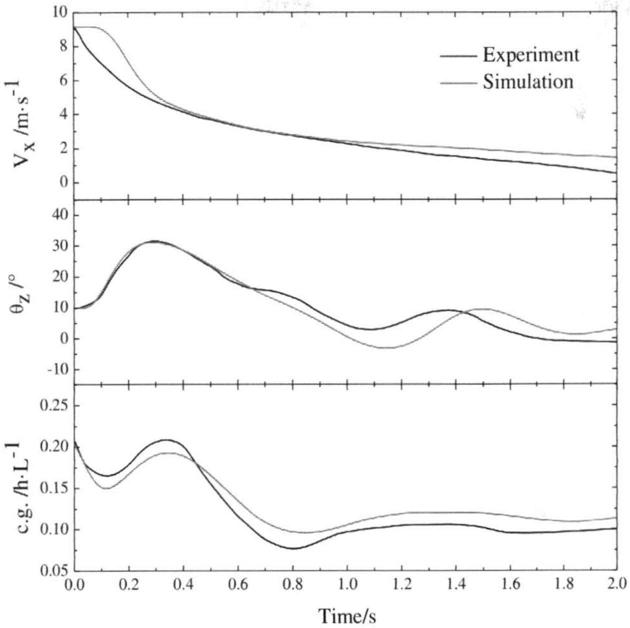

Figure. 1 Comparison between the experimental and numerical results for model F

3. Physical Model and Wave Model

3.1. Physical Model and Computational Grid

The model studied in this paper is a high wing high tail regional aircraft with normal configuration, which is basically the same as the model F of NACA TN 2929 model. A half mesh of the model is used and the aircraft is located at the centre of the cube computational domain. The boundary is set at 5L (L is the body length) away from the model. A mix of unstructured and structure mesh of the total cell number of 3.8 million is used, as Figure 2 shows.

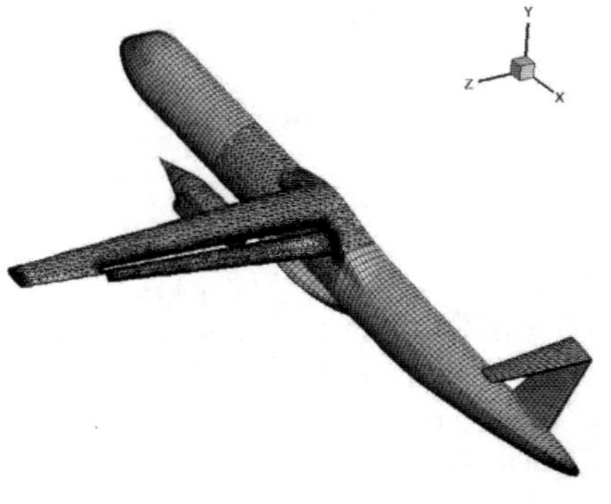

Figure. 2 Surface mesh layout of the model in this article

3.2. Wave model

Water surface is rough with swell in most cases. In this section, a sinusoidal curve is used to represent the swell. Wavy water ditching can be separated into to two characteristic situations: vertical to swell and parallel to swell. Ditching parallel to the swell approaches calm water ditching. Ditching vertical to swell is more dangerous when an aircraft flies against the wave propagation direction. The impact load is related to the vertical relative velocity between the aircraft and water surface. In order to study the extreme impact load during ditching, ditching vertical to swell is selected for research.

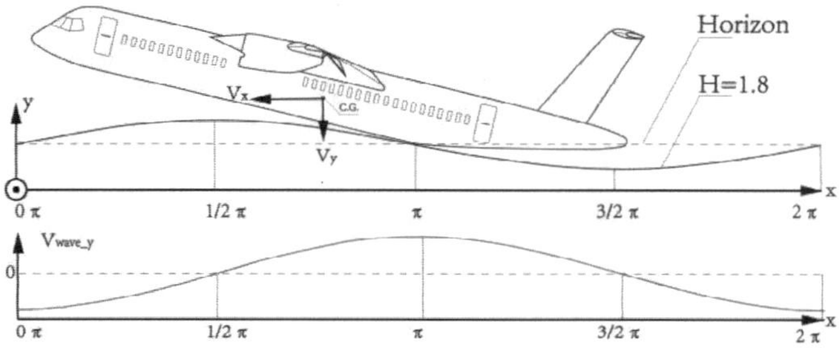

Figure. 3 Wave surface shape and vertical velocity distribution

As shown in Figure 3, the aircraft moves from right to left. The ground coordinate system is taken as a reference, the X axis is positive to the right, the y axis is positive upwards, and the pitch angle θ_z of the aircraft is positive to nose up. A sinusoidal wave with small amplitude is selected in the calculation.

$$\Delta h = \frac{H}{2}\sin\left(\frac{2\pi}{L}x - \frac{2\pi}{T}t\right) \tag{1}$$

$$v_y = -\frac{\pi H}{T}\cos\left(\frac{2\pi}{L}x - \frac{2\pi}{T}t\right) \tag{2}$$

Where Δh is the position of wave surface, and v_y is the vertical velocity of wave. In this paper, the wavelength is 37.5 m, the wave height is 0.6m to1.8m, and the period is 4.9 s. The above conditions belong to level 3 to 4 sea conditions.

Von Karman [4] shows that the impact load is proportional to the square of the impact velocity. In this paper, the most dangerous landing condition is located at the phase π of the water surface where the vertical relative velocity between the aircraft and water surface is the largest, as shown in Figure 3.

4. Analysis of Typical Ditching Process in Wave Environment

According to the parameter range recommended by ditching procedure [2], the landing velocity is 1m/s, pitch attitude is 10.79° (the angle of attack is 12°), the airspeed of the aircraft is 47.5m/s, and the sideslip angle is 0°. At the beginning of the cases, the lower surface of the fuselage is 0.2m away from the water surface. In the following analysis, the load coefficient is calculated by Equation(3) and (4):

$$F_x = \frac{f_x}{mg} \tag{3}$$

$$F_y = \frac{f_y}{mg} \tag{4}$$

Where F_x is the horizontal load, F_y is the vertical load, f_x is the horizontal force, f_y is the vertical force. Figure 4 shows comprehensive curves of aircraft ditching on calm surface and wavy surface, where α_z is pitch angle acceleration. According to the curve, the ditching process can be divided into the following three stages: 1, the contact stage, which occurs in 0 to 0.3s. The fuselage just touches the water surface. 2, the impact stage, which occurs in 0.3 to 4s. The fuselage impacted into the water and accompanied with planing. The change of load is violent in this stage. 3, the planing stage. After 4s the planing in horizontal direction is dominant at the end of the ditching behavior. The stress situation of the aircraft in this stage tends to be stable, and there is no more load peak.

In the contact stage, there are obvious difference between the calm surface and the wavy surface ditching. Figure 5 shows the comparison of the aircraft attitude and water surface shape in the contact phase. Figure 6 shows the comparison of the aircraft ditching load process in this phase.

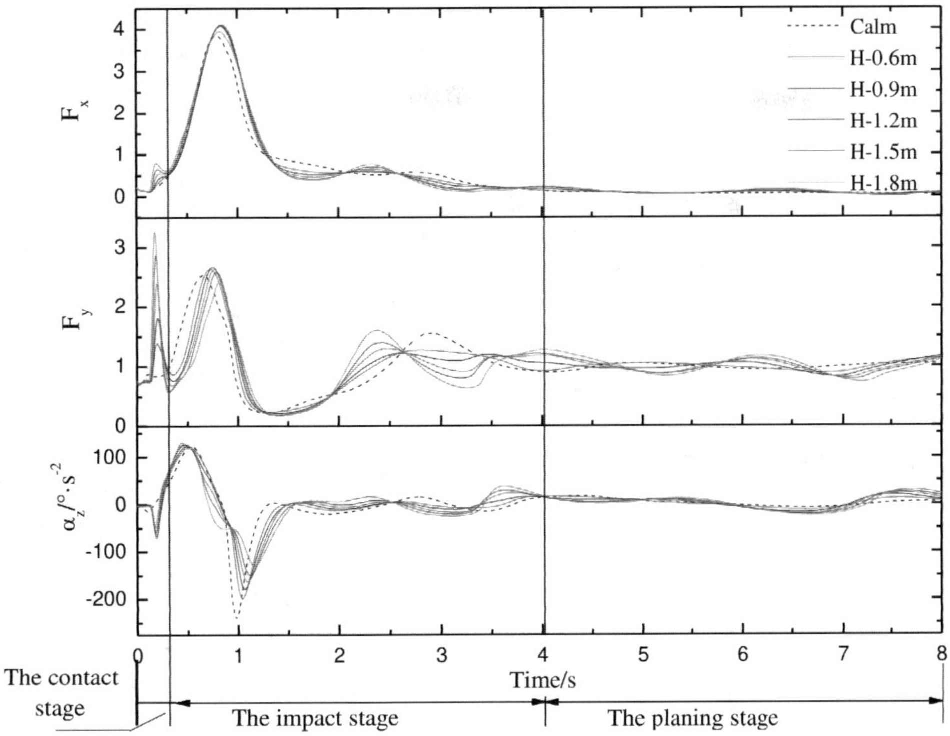

Figure. 4 Load history of ditching on calm and wave surface

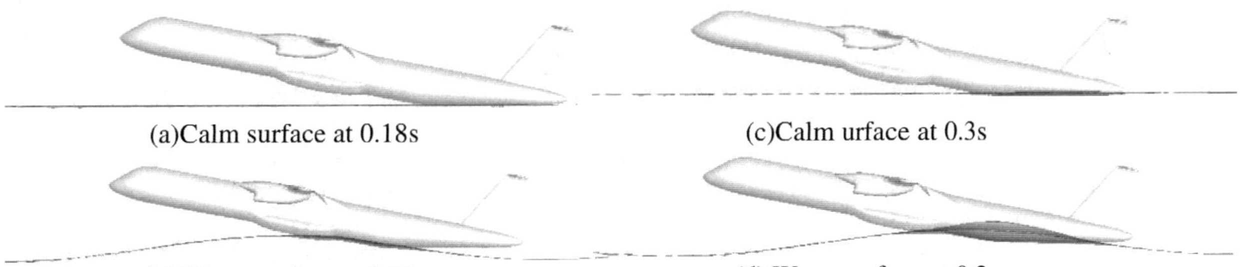

(a)Calm surface at 0.18s (c)Calm urface at 0.3s

(b) Wave surface at 0.18s (d) Wave surface at 0.3s

Figure. 5 Aircraft attitude and water surface during water contact stage

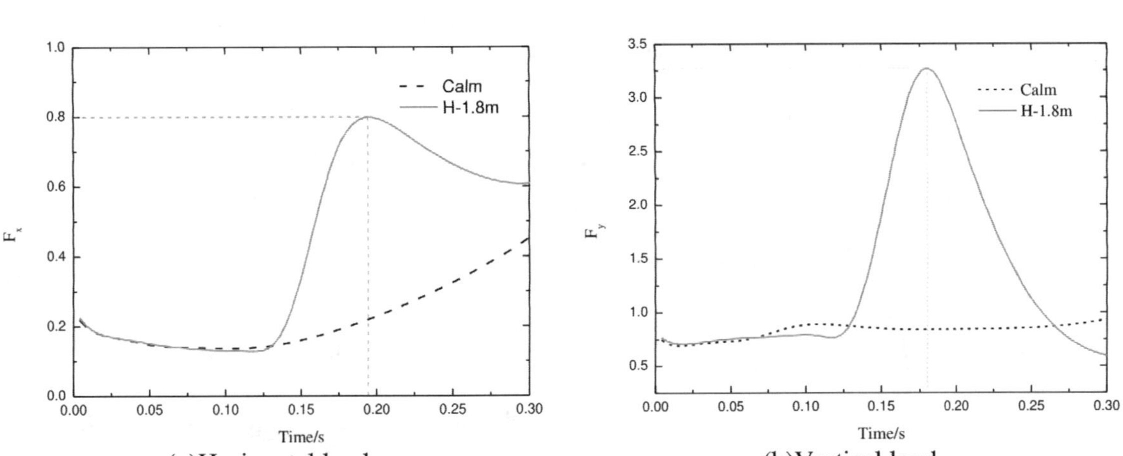

(a)Horizontal load (b)Vertical load

Figure. 6 Motion and load history during water touch stage

Figure 6(a)(b) shows ditching in calm surface. The rear of the fuselage evenly contacts the water surface, showing a planing attitude. As can be seen from Figure 8, the planing behavior leads to an increase in horizontal drag. At this moment, the depth of the fuselage tail in water is shallow. The ditching behaviour has not yet developed into the impact process, so the vertical load increase is very weak. The aircraft continues to plan underwater and gradually transits to the impact stage.

As can be seen from Figure 5(c) and 6, when the abdomen of the fuselage touches the water facing the wave slope the rear of the fuselage has not yet touched the water. A small drag peak appears in the horizontal direction at 0.19 s during the fuselage penetrating the wave crest, and the horizontal load reaches 0.80. At this time, the vertical relative velocity between the aircraft and water surface is the largest. A big impact load peak appears and the vertical load is as high as 3. 27 when the fuselage penetrates the wave.

The above contents confirm the hypothesis of this paper that ditching in wavy surface may have unexpected impact load peak due to the different position of water contact.

5. The Characteristics of Ditching in Different Wave Heights

As shown in Figure 7, that oscillation velocity of the water surface are different for different wave-forms. According to equation (2), the oscillation velocity of the water surface is proportional to the wave height at the same phase.

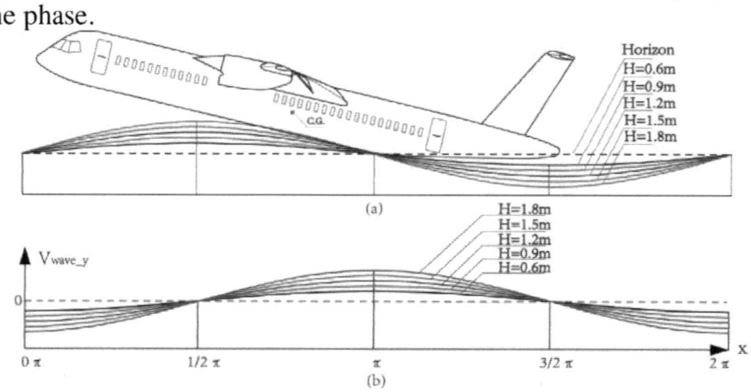

Figure. 7 Wave surface and velocity distribution in different wave heights

In this section, the simulation results of ditching in wavy surface in five different wave heights are compared with the result in calm surface, and the corresponding peak values are extracted for lateral comparison under different wave heights.

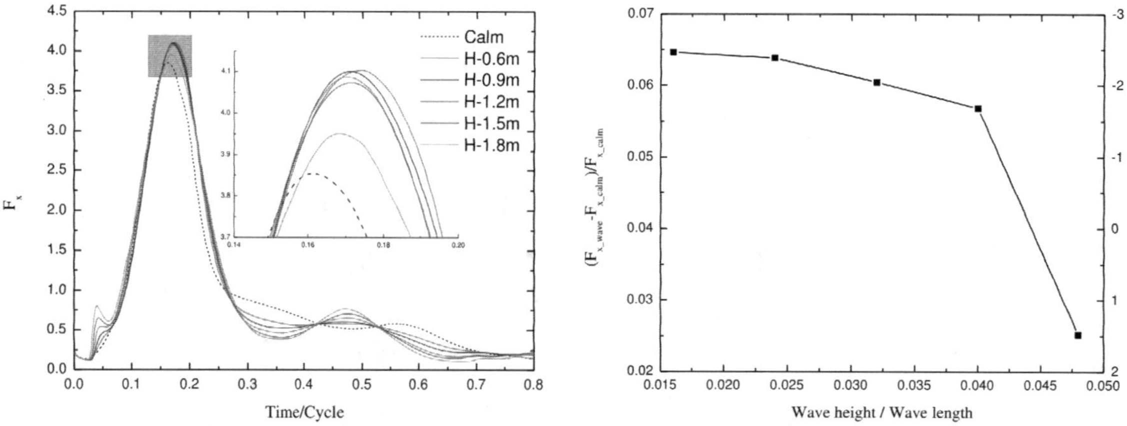

(a) Horizontal load history (b) Horizontal load peak comparison

Figure. 8 Horizontal load comparison

Figure 8(a) shows the peak of the horizontal load in the impact stage, and the peak of the horizontal load in the wave surface is greater than that in the calm surface. The differences between them is small. As shown in Figure 8(b), with the increase of the wave height, the increase of the peak relative to the calm surface is smaller. When the wave height/wavelength is 0.016, the peak of the horizontal load will be greater than 6.5% of the calm surface. This is because as the wave height increases, the peak of horizontal load in the contact stage increases, the horizontal velocity decreases, so the peak of horizontal load in the impact stage decreases.

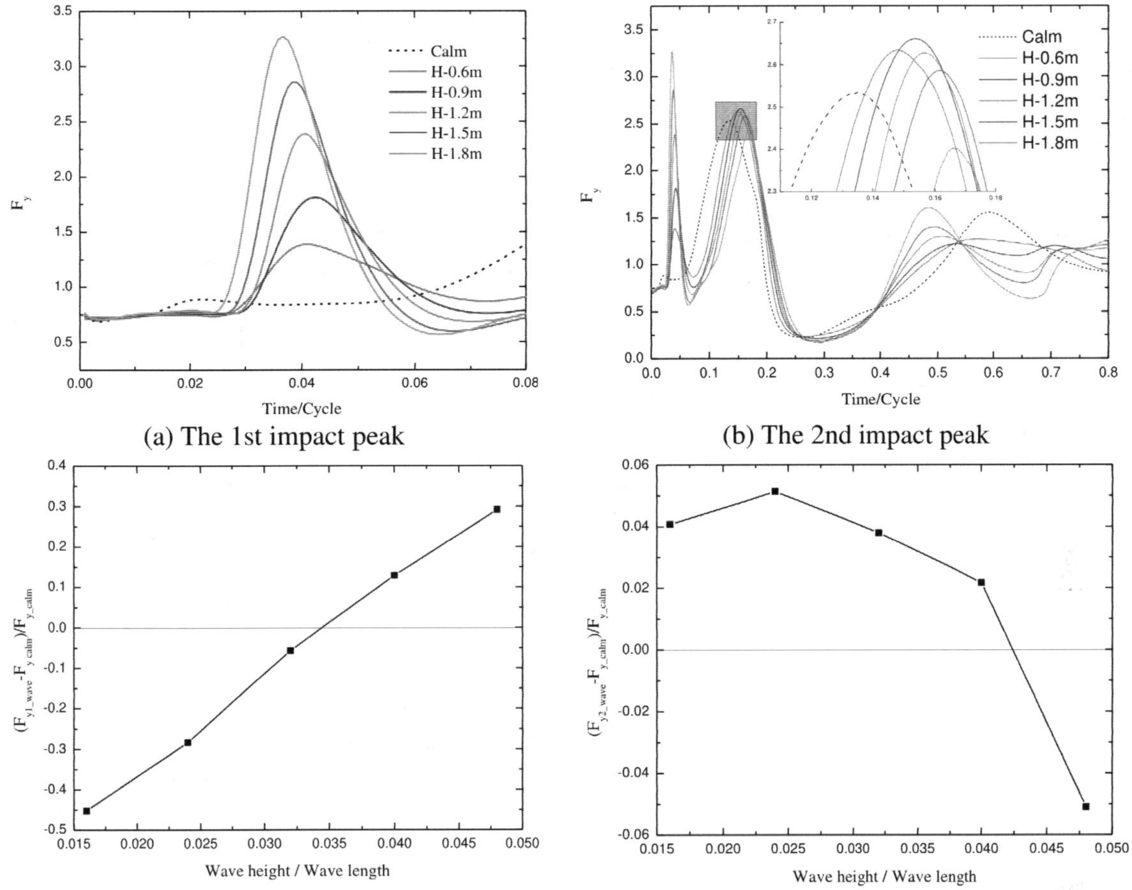

(a) The 1st impact peak (b) The 2nd impact peak

(c) The 1st impact peak comparison (d) The 2nd impact peak comparison

Figure. 9 Vertical load comparison

Two vertical load peaks are in the wavy surface. The first one is in the contact stage and the second one is in the impact stage. Both of them are compared with the only one vertical load peak in the calm water surface. The differences among the first impact peaks in the wavy surface are obvious. It is because the landing velocity of the aircraft is small, so the influence of the water surface oscillation velocity is obvious in the vertical relative velocity between the aircraft and water surface. The vertical impact peak value increases obviously with the increase of the wave height. Figure 9(c)(d) shows that with the increase of the wave height, the first impact peak value increases, and the second impact peak value basically decreases. This is consistent with general cognition. It is known that there is at least one vertical impact peak value greater than that in calm surface for any wave height. When the wave height/wavelength is 0.048, the extreme impact load peak value will exceed 30% of that in calm surface.

6. Conclusion

In this paper, Ansys Fluent software combined with the GMM is used to simulate the ditching process of an aircraft in five different wave heights. The results are compared with that in calm surface. In order to study the extreme load during the wavy surface ditching, this paper chooses the π phase as the landing point. When the aircraft approaches the water surface with the recommended landing attitude, the landing point is the abdomen of the fuselage, which is different from the rear of the fuselage in the calm surface. Therefore, there is an unexpected impact peak in the wavy surface in the contact stage. Through the parametric analysis, it is found that with the increase of wave height, the impact peak value in the contact stage increases gradually. When the wave height is large enough, the vertical impact peak value in this stage will far exceed the vertical load of the calm surface and become the extreme load in the whole ditching process. In the impact stage, the horizontal load peak value in the wavy surface is slightly larger than that in the calm surface. With the increase of wave height, the vertical impact peak value decreases gradually. However, during the whole ditching process, at least one of the two vertical impact peaks in the wavy surface is larger than that in the calm surface.

7. References

[1] NTSB. Revised 12-98, Aviation Coding Manual[S], NTSB: Washington DC, 1998.
[2] PATEL A A, GREENWOOD R P Jr. Transport water impact and ditching performance, Final Report DOT/ FAA/ AR-95/54 [R]. US Department of Transportation: WASHINGTON DC, 1996.
[3] THOMSON R G，CAIAFA C. Structural Response of Transport Airplanes In Crash Situations, DOT/FAA/CT83/42[R]. NASA: United States, 1983.
[4] VON KARMAN T. The impact of seaplane floats during landing, NACA TN 321[R]. NACA: Washington DC, 1929.
[5] WAGNER H. Phenomena associated with impacts and sliding on liquid surfaces[J]. Journal of Applied Mechanics, 1932, 12: 193-215.
[6] MCBRIDE E E, FISHER L J. Experimental investigation of the effect of rear-fuselage shape on ditching behavior, TN 2929[R]. Washington DC, NACA: 1953.
[7] XU G D, DUAN W Y, WU G X. Numerical simulation of water entry of a cone in free-fall motion[J]. Quarterly Journal of Mechanics and Applied Mathematics, 2011, 64: 265-285.
[8] HUA C, FANG C, CHENG J. Simulation of fluid-solid interaction on water ditching of an airplane by ALE method[J]. Journal of Hydrodynamics, 2011, 23(5):637-642.
[9] STRECKWALL H, LINDENAU O, BENSCH L. Aircraft ditching: a free surface / free motion problem[J]. Archives of Civil and Mechanical Engineering, 2007, 7(3):177-190.
[10] QU Q L, HU M X, GUO H, et al. Study of Ditching Characteristics of Transport Aircraft by Global Moving Mesh Method[J] Journal of Aircraft, Volume 52, Issue 5, 2015, Pages 1550-1558.

IOP Conf. Series: Materials Science and Engineering **816** (2020) 012004 doi:10.1088/1757-899X/816/1/012004

Numerical Simulation of the Aerodynamic Performance of the Backswept Blade of a Propeller at High Altitudes

J M Chen[1], G Y He[1], Q Wang[1] and L S Chen[1]

[1] Nanchang Hangkong University, Nanchang 330063, China

HEGY509@ 163.com

Abstract. In high-speed situations, propeller efficiency drops rapidly after the design point is surpassed, and propeller performance deteriorates sharply; thus, high-altitude, long-endurance unmanned aerial vehicles (UAVs) cannot readily meet the design requirements of fast climbing and rapid maneuvering. The design method of the swept blade is investigated to solve this problem. The three-dimensional Reynolds averaged Navier-Stokes (RANS) equations are solved based on the periodic boundary conditions and multiple reference frame (MRF) models to simulate the aerodynamic properties of unswept and swept blades. The results show that the swept-blade propeller can provide a larger thrust than the unswept-blade propeller and that the velocity range with acceptable efficiency is extended after the design point is surpassed. The sweep design method used in this paper provides a reference for the propeller design of high-altitude, long-endurance UAVs.

1. Introduction

Research on high-altitude, long-endurance UAVs has become a hot topic in modern aviation. Unlike other low-altitude aircraft, due to their long life, long endurance and diversity of missions, high-altitude, long-endurance UAVs have broad prospects in both the military and the civilian fields[1-3]. All of these characteristics rely on a high-efficiency propulsion plant. Typically, a propeller is the best choice for high-altitude, long-endurance UAVs and airships, with an efficiency unmatched by that of jet engines[4]. However, the air environment at the stratosphere (approximately 20 km), where high-altitude, long-endurance UAVs fly, is different from that at low altitudes (approximately 10 km)[5]. In particular, the propeller faces the problem that the airflow bypasses the blade easily when working under a low Reynolds number (10^4~10^5), which can have a negative impact on the efficiency and aerodynamic performance of the propeller compared to the response at low altitudes. Moreover, the thin air at high altitudes is not conducive to heat dissipation. Therefore, a fixed-range propeller with a simple transmission structure has been adopted to solve these problems. In short, the aerodynamic efficiency of the propeller has a direct and significant impact on the performance of high-altitude, long-endurance UAVs, including the fuel economy, climb performance, and operating conditions.

In this article, the swept propeller blade is studied. Under the premise of adding no additional mechanical devices or control systems to the propeller, this paper applies the CFD (Computational Fluid Dynamics) numerical method to simulate the viscous flow around a swept-blade propeller and study the swept-back design of the propeller blades. Furthermore, the effects of different sweep positions and sweep angles on the performance of high-altitude propellers are summarized to provide a reference for the design and improvement of propellers suitable for high-altitude, long-endurance UAVs.

2. Specification of the Blade Geometry Model

The radius of the baseline propeller is 2.3 m, and the blade element features the Eppler387 airfoil. Using the Z-axis in Cartesian coordinates as the stacking line and the midpoint of the airfoil chord line located at the Z-axis, 10 blade elements are evenly distributed along the spanwise path. The distributions of the chord length and torsion angle are provided in reference[6], and the one of the unswept blades of baseline propeller is shown in Figure 1. The swept-blade model of the propeller of interest is based on the structural elements of the unswept blade of baseline propeller. Based on the unswept blade of baseline propeller, the following changes are applied to obtain the swept blade. First, the chord line of each blade element of blade of baseline propeller is extended to form a torsion surface consistent with the twist angle of the unswept blade. Second, the stacking line of the unswept blade is divided at the position r/R into two parts. The upper part of the stacking line at α degrees is swept along the arbitrary direction, the upper stacking line is extended to make the Z coordinates of the endpoint equal to those of the unswept-blade stacking line, and then the broken line is rotated around the Z-axis at an arbitrary angle so that the rotating surface intersects with the torsion surface. Using the endpoint of the intersection line, the midpoint and starting point of the stacking line are used to generate a spline curve, and then this curve is projected to the torsion surface to form the stacking line of the swept blade. Finally, the chord length and torsion angle distribution of the unswept blade are used to form a blade with a sweep angle of α degrees at r/R of the blade spread. The swept-blade modeling method is shown in Figure 2. Two parameters, the sweep angle α and the sweep position r/R, are adopted to parametrize the swept propeller, as shown in Table 2.

Figure 1. Unswept-blade model.

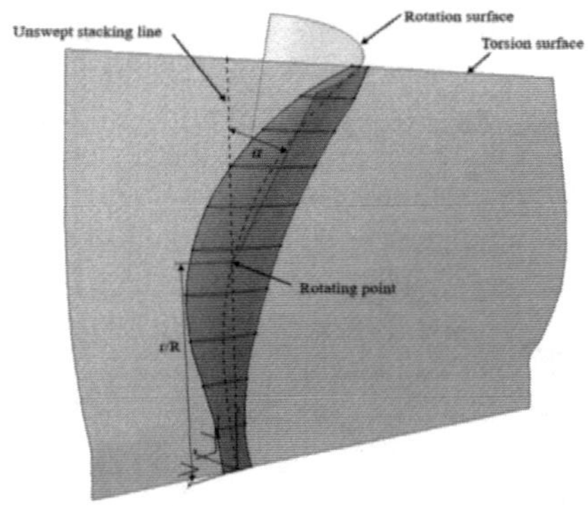

Figure 2. Swept-blade model.

Table 1. The two parameters of interest for the swept blade

sweep position	Sweep angle (°)
0	10;30
0.1	10;30
0.2	10;30;40
0.3	10;30;40
0.4	10;30;40;50
0.5	10;30;40;50
0.6	10;30;40;50
0.7	10;30;40;50
0.8	10;30;40;50
0.9	10;30;40;50

IOP Conf. Series: Materials Science and Engineering **816** (2020) 012004 doi:10.1088/1757-899X/816/1/012004

3. Numerical Simulation Method and Mesh Independence Verification

The propeller is working at a height of 25.9km, the atmospheric physical properties are as follows: the atmospheric pressure P=2549.2pa, the density ρ=0.03475kg/m^3, the viscosity μ=1.46 × 10-5kg/(m·s), the temperature T=221.5K, the thermal conductivity Cp=1013J/ (kg·k). The numerical simulation is based on Cartesian mesh and periodic boundary condition, thus, only a single blade is calculated. The computational domain is divided into two regions, the rotational and stationary zones as presented in Figure 3. The boundary of stationary zone were defined with a pressure far field boundary condition. The y+ was controlled by the external refinement of the mesh near the blade, through the creation of hexahedral boundary cells around the surface of the blade such that the first point above the blade surface had a value y+ <1. The cells around the blade is shown in Figure 4. The Reynolds averaged Navier-Stokes(RANS) equations and k-ω SST turbulence model with Low-Reynolds modification are solved by multiple reference frames(MRF) model and the implicit coupled density based solver is used, the spatial discretization of the flow convective fluxes and turbulence was performed by high order advection upstream splitting method(AUSM) schemes and second order upwind scheme respectively. We set the mesh size of the blade surface to 10 mm, 7 mm, 5 mm and 3 mm, corresponding to volume element numbers of 10.75 million, 11 million, 13.25 million and 13.9 million, respectively. Mesh independency tests were performed to ensure that the obtained results were not dependent on the mesh used. As presented in Figure 5, when the blade surface mesh size is less than 5 mm (body mesh is 13.25 million), the fluctuation of the calculated pull and moment is very small. Therefore, a blade surface mesh of 5 mm is used to investigate the unswept blade and swept blade.

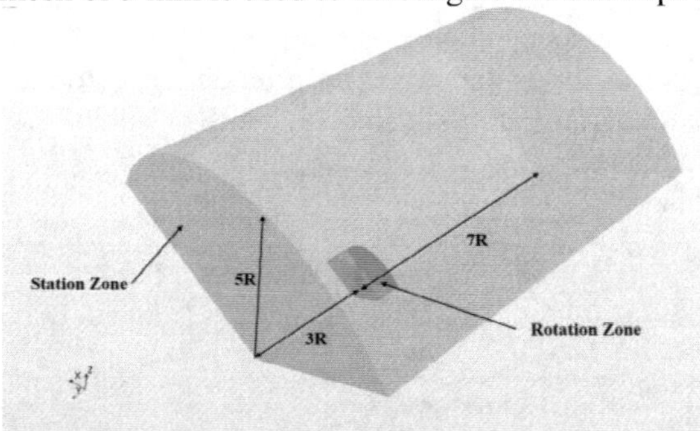

Figure 3. Computational domain.

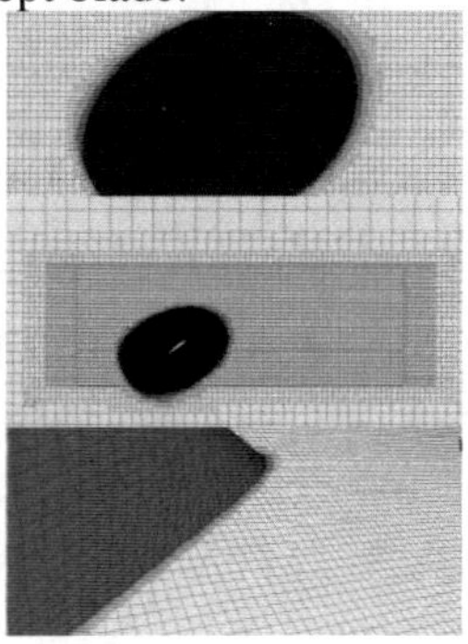

Figure 4. Meshing around the blade.

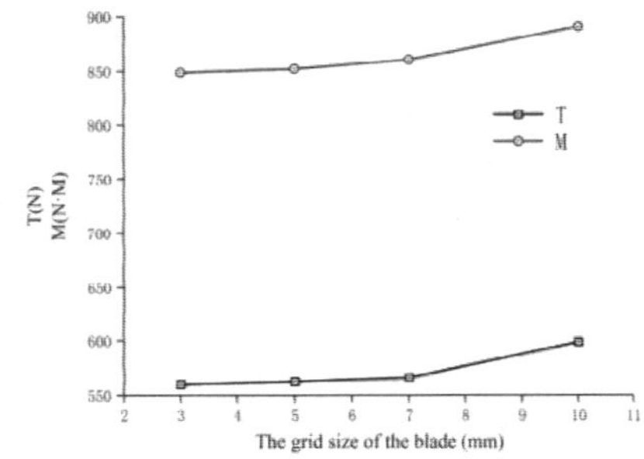

Figure 5. Mesh independency study.

4. Results and Discussions

Figures 6, and 7, respectively show comparisons of the thrust and efficiency of swept-blade propellers with the same sweep angle at different sweep positions. Overall, the trends of thrust and efficiency are similar. With increasing flight speed of the high-altitude, long-endurance UAVs, the thrust generated by the propeller is linearly reduced. The efficiency of the unswept-blade propellers is higher than that of the swept-blade propellers when the flight speed is lower (for advance ratios of 1.7-1.9). As the flight speed increases, the efficiency of the swept-blade propellers grows much higher than that of the unswept-blade propellers.

The case of a small sweep angle (α=10°) of the swept-blade propeller is examined first. Figure 6(a) shows that as the sweep position gradually moves toward the paddle, the thrust of the propeller increase gradually, and the maximum influence of the sweep position on the thrust and power is less than 1%. In combination with the efficiency comparison curve of Figure 7(a), it can be concluded that the maximum difference in efficiency between the swept-blade propellers and the baseline propeller is also less than 1%. When the advance ratio is within the range of 1.7 to 2.0, the influence of the sweep position on the propeller efficiency is less than 1%, and the maximum peak efficiency appears at J=2.0. The advance ratio increases to 2.3, and the change in the sweep position has a stronger effect on the efficiency. However, the efficiency of the swept-blade propellers under these conditions is reduced by more than 15% compared to the peak efficiency at an advance ratio of 2.0, and the baseline propeller has sharp decreasing. The case of sweep angle is greater than 30° of the swept-blade propellers are examined next. Figure 6, (b),(c),(d) show that when the pitch ratio is 1.7, the thrust generated by the swept-blade propeller first at the maximum at a sweep position of r/R=0.5 and then gradually decreases as the blade sweep position r/R changes. As the flight speed continues to increase (J>1.7), the thrust generated by the swept-blade propeller gradually increases as the sweep position of the blade moves closer to the root of the blade (corresponding to a smaller r/R). As shown in Figure 7(c), as the advance ratio is to the range of 1.7-1.9, the closer the sweep position is to the blade tip (corresponding to a larger r/R) and the more efficient the swept-blade propeller is. An advance ratio of 2.1 corresponds to the peak efficiency. As the flight speed continues to increase, the efficiency of the swept-blade propeller first increases and then decreases as the sweep position changes from 0.5 to 0.9. The efficiency of the swept propeller at an advance ratio of 2.3 is 15% higher than that of the baseline propeller.

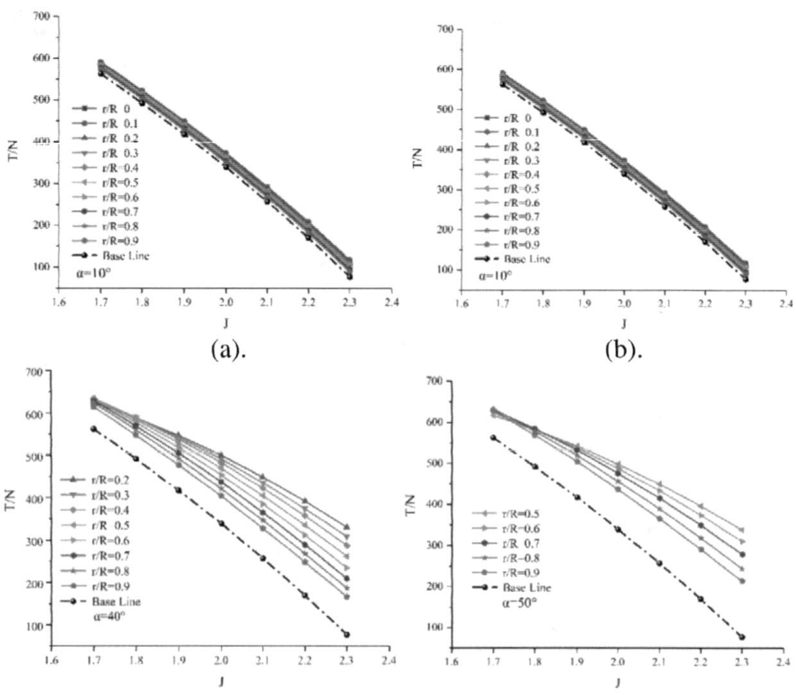

(a). (b).

(c). (d).

Figure 6. Thrust comparison of different swept-blade propellers.

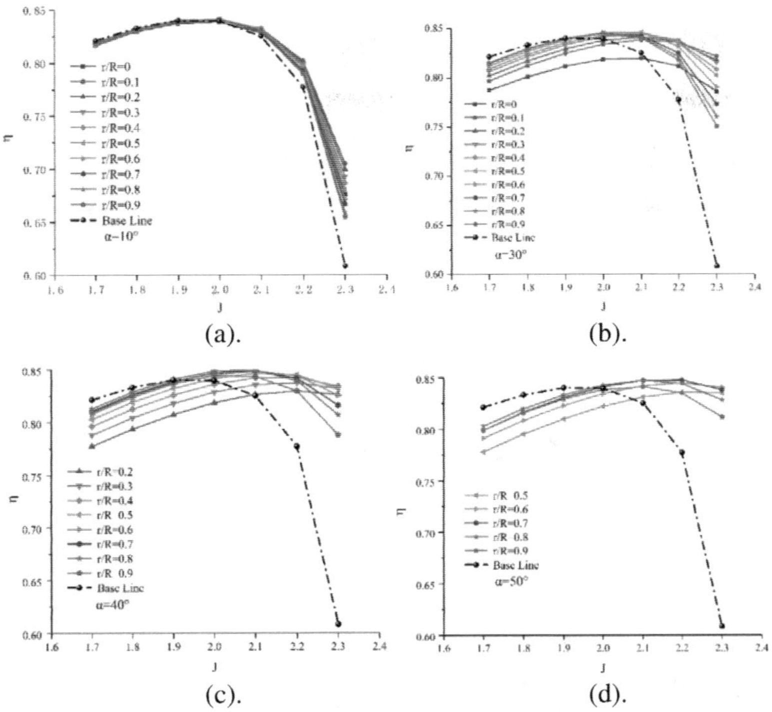

Figure 7. Efficiency comparison of different swept-blade propellers.

5. References

In summary, changes in the sweep angle and sweep position affect the propeller performance. According to the above description, the following conclusions can be drawn:

1. Compared to the baseline propeller, the blade propeller with a small sweep angle (α=10°) brings less than a 1% increase in thrust, power and efficiency. Therefore, a small-angle sweep of the propeller blade (α=10°) is not a good choice for high-altitude, long-endurance UAVs;
2. The propeller efficiency at high speed starts to improve substantially starting at a sweep angle of 30°. The best sweep angle is between 30° and 50°. Under these conditions, the efficiency of the propeller in the lower speed range does not change much, and this setting can substantially improve the efficiency of the propeller in the higher speed range and effectively widen the working range of the propeller.

6. References

[1] Colozza A and Dolce J 2003 Initial feasibility assessment of a high altitude long endurance airship
[2] Jamison L Sommer G S and Porche III I R 2005 High-altitude airships for the future force armyed (Rand Arroyo Center Santa Monica CA)
[3] Moomey E R 2005 Technical feasibility of loitering lighter-than-air near-space maneuvering vehiclesed
[4] Carichner G E and Nicolai L M 2013 Fundamentals of Aircraft and Airship Design Vol 2– Airship Design and Case Studies (American Institute of Aeronautics and Astronautics Press)
[5] Tang Z Liu P Chen Y and Guo H 2015 Experimental study of counter-rotating propellers for high-altitude airships J. Journal of Propulsion Power. pp 1491-1496
[6] Koch L D J N 1998 Design and performance calculations of a propeller for very high-altitude flight

IOP Conf. Series: Materials Science and Engineering **816** (2020) 012005 doi:10.1088/1757-899X/816/1/012005

Numerical Study on the Aerodynamic Performance of the Rigid and Corrugated Forewing of Dragonfly in Flapping Flight

H Song[1], G Y He[1], Q Wang[1], L S Chen[1]

[1]Nanchang Hangkong University, Nanchang 330063, China

HEGY509@ 163.com

Abstract. The dragonfly flapping flight as a kind of important flight model, in order to explore the aerodynamic effect on the forewing of dragonfly in flapping. Based on the real dragonfly wings(with corrugation 3D forewing), this paper uses X-flow software numerical simulation for the model, analyzed the influence of different flow velocity and frequency on the aerodynamic effect of the forewing of dragonfly. When the dragonfly flies forward stably, the results show that it has a higher lift with the increase of inflow velocity and flapping frequency. The leading edge vortex (LEV) of the dragonfly's forewing does not fall off in a flapping period, and the maximum positive lift is generated in the downstroke. When the trailing edge vortex is strong, the peak value of thrust coefficient appears in the process of flapping, and the energy of the trailing edge vortex may provide the thrust for the dragonfly forward flight.

1. Introduction

Dragonflies, as the most capable fliers in nature, exhibit a variety of flight modes, including glide flight, flapping wing forward flight, tandem flight, low-speed maneuver and hovering, and they have been recorded to fly at speeds of up to 10m/s[1,2]. The excellent flying ability of dragonflies mainly comes from the ingenious structure of wings and a variety of flapping ways[3]. The dragonfly has high mobility, good controllability and low noise level, which makes it an ideal choice for bionic micro air vehicles (MAVs), so it has important application value in military investigation, search and rescue[5].

At present, there are many researches on the aerodynamic characteristics of dragonfly wings around the world. Kesel's analysis of dragonflies cross-section folds shows that folds can improve the lift of dragonflies wings[6]. Sun and Lan demonstrated that the interaction between the forewing and hindwing can reduce the vertical force compared to the single moving wing[7].

The model in this paper refers to the model built by Liu. According to the actual structural characteristics of dragonfly's forewing and the measurement of dragonfly's forewing, ignoring the extremely small corrugated and the influence of corrugated internal structure, as well as the slight change of wing vein shape, and simplifying the wing vein shape of dragonfly's forewing into a circular tube, as shown in figure 1. X-Flow software was used to simulate the model numerically in the case of unsteady flow field with low Reynolds number. Solving the flow field characteristics around the forewing of dragonflies. Analyze the impact of flapping frequency and inflow velocity on the aerodynamic efficiency of dragonfly flapping. I hope to provide necessary theoretical and technical support for the development and improvement of the flapping aircraft.

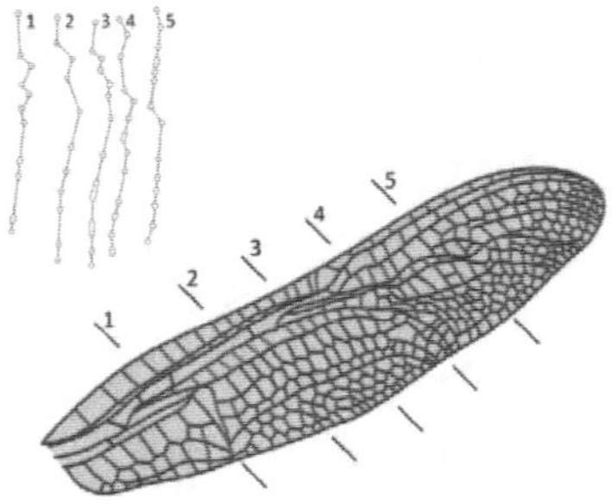

Figure 1. 3D corrugated forewing model of dragonfly

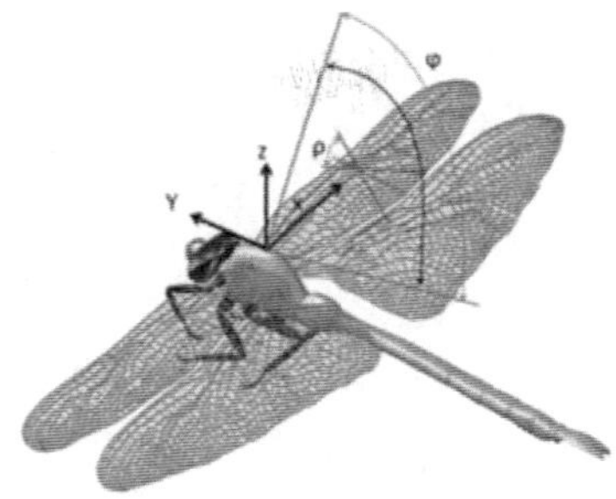

Figure 2. Schematic diagram forewing of dragonfly movement

2. Physical models and numerical methods

2.1. Flapping model

In order to conveniently show the position of the forewing in space and the flapping parameters of the forewing when the dragonfly flapping, the forewing of the dragonfly is assembled on the body of the dragonfly, as shown in Figure 2. But for reasons of simplified numerical simulation, this paper ignores the dragonfly body and the influence of the hindwing on the aerodynamic efficiency of the forewing, only numerical simulation of corrugated forewing of dragonfly. When the flapping wing of dragonfly forward flight, the forewing is doing complex space movement, which can be simplified as the overlapping of two parts of flapping up and down around the body axis (z axis) and twisting around the wing span direction (x axis).

2.2. Flapping equation

The equation of flapping Angle a of dragonfly's forewing changing with time t is

$$\varphi = \varphi_0 + A\cos(2\pi ft - \theta_0) \quad (1)$$

The equation of torsion Angle a of dragonfly forewing changing with time t is

$$\rho = \rho_0 + \sum_{n=1}^{3} B_n \cos(2\pi fnt + \theta_n) \quad (2)$$

2.3. Setting Project

In order to study the influence of different inlet velocity and frequency on the flow field and aerodynamic force around the forewing of dragonfly, this paper carried out numerical analysis on the three-dimensional corrugated forewing model of dragonfly's with 5 groups of different inlet velocity and 5 groups of different frequencies. According to Wang's[8]measurement, the default forward flight speed is 1.71m/s, and the frequency is 33.4Hz.

3. Numerical simulation and Result analysis

3.1. Influence of inlet velocity on aerodynamic characteristics of flapping wing

Fig 3 and Fig 4 show the curves of lift coefficients and drag coefficients with time in one cycle. It can be seen from the figure that the curves of lift coefficient and drag coefficient change the same trend at different speeds. At the beginning, the forewing of dragonfly upstroke from the lowest point. When forewing upstroke, the lift coefficient is negative; when forewing downstroke, the lift coefficient is positive. The maximum value of negative lift is generated in the upstroke, and the maximum value of positive lift is generated in the downstroke. The value of lift coefficient decreases with the increase of velocity. When the

IOP Conf. Series: Materials Science and Engineering **816** (2020) 012005 doi:10.1088/1757-899X/816/1/012005

speed is small, there will be drag at the beginning of flapping. When the velocity is greater than or equal to 5m/s, the whole flapping period of the forewing is almost negative, that is, the thrust is generated, and the higher the velocity, the smaller the thrust is generated. In a flapping cycle, the thrust produces two peaks,one in the upstroke and one in the downstroke, among which the thrust peak generated in the downstroke is about 2 times of the upstroke.

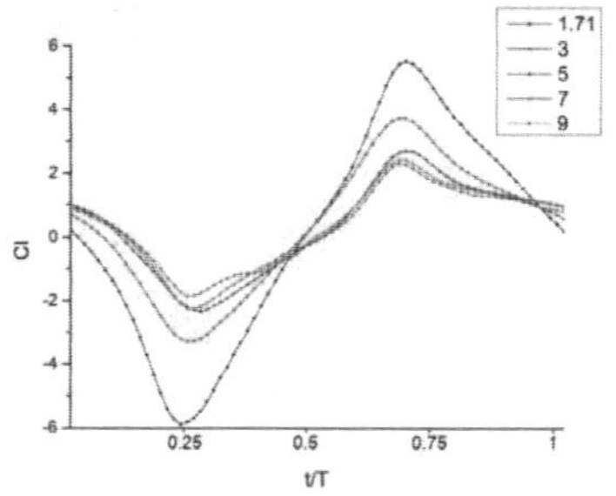

Figure 3. Curve of lift coefficient with time in a cycle

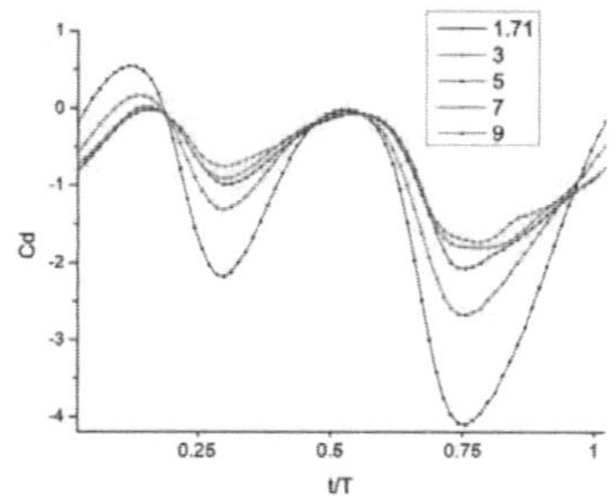

Figure 4. Curve of drag coefficient with time in a cycle

Figure 5 shows the pressure distribution on the upper and lower surfaces of the forewing at different times when the velocity is 5m/s. In the initial position, the forewing of the dragonfly starts to upstroke from the lowest point, and the pressure difference between upper and lower surfaces gradually increases. When it reaches 0.25T, the dragonfly upstroke the horizontal position, the pressure on the upper surface is significantly higher than that on the lower surface. The pressure difference between the upper and lower surfaces reaches the maximum, generating the maximum negative lift, which corresponds to the trough value of the lift coefficient curve(in Figure 3). When the forewing continues to upstroke to the highest point(0.5T), the differential pressure between the upper and lower surfaces is almost zero, and the drag coefficient generated is almost zero. After reaching the highest point, the forewing of the dragonfly starts to downstroke. During the downstroke, the flapping speed increases and the pressure difference between the upper and lower surfaces gradually increases. At 0.75T, a low pressure area is formed on the upper surface and a high pressure area is formed on the lower surface. The pressure difference between the upper and lower surfaces reaches the maximum, generating the maximum positive lift, which corresponds to the peak value of the lift coefficient curve (Figure 3). As the wing tilts forward at this time, the thrust coefficient also reaches the peak value (Figure 4).

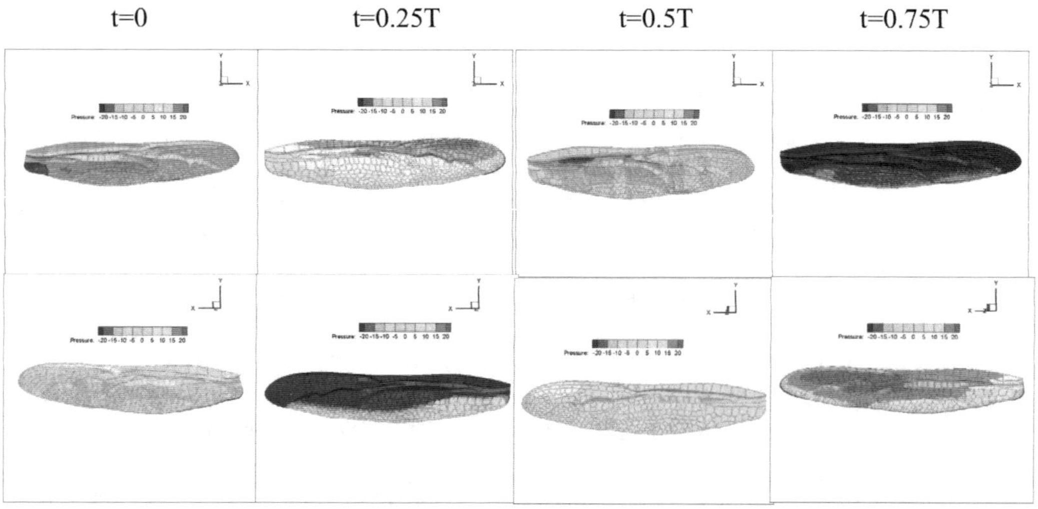

Figure 5. Upper and lower surface pressure cloud(v=5m/s)

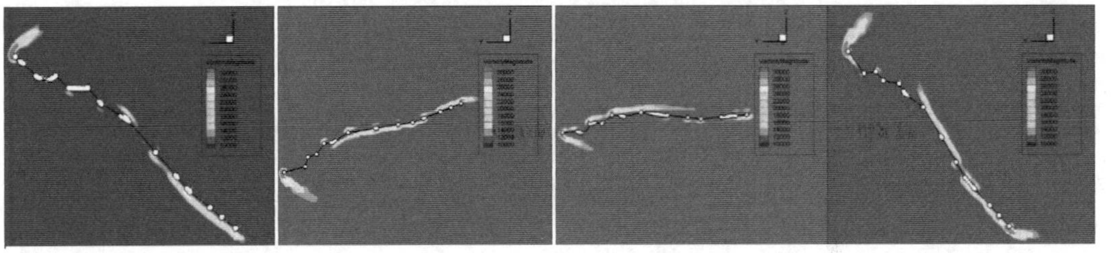

t=0 t=0.25T t=0.5T t=0.75T

Figure 6. Vorticity of forewing cross sections show at 0.5L(where L is the span length)

When t= 0.25T, the forewing of dragonfly begin to upstroke from the lowest point, the vorticity mainly appears in the lower wing surface, and also produces a large trailing edge vortex. At this time, the maximum negative lift and the first thrust peak appear corresponding to figure 3 and figure 4. Then, the forewing continues to upstroke. When t= 0.5t, the forewing upstroke to the highest point, at this time, the trailing edge vortex is very small and the thrust coefficient is almost zero. Next, the downstroke starts from the highest point. When t= 0.75t, vorticity appears on the upper wing surface and generates strong trailing edge vortex. At this time, the maximum value of positive lift coefficient and the peak value of thrust coefficient appear for the second time (corresponding to figure 3 and figure 4). The lift coefficient C_L and drag coefficient C_D in the flapping of dragonfly come from the non-stall mechanism (leading edge vortex does not fall off), and the generation of trailing edge vortex provides thrust for the flapping of dragonfly.

3.2. Influence of frequency on the aerodynamic characteristics of flapping

As can be seen from figure 7 and figure 8, with the increase of frequency, the variation trend of lift coefficient and drag coefficient is the same. With the frequency increases, the corresponding values of wave crest and wave trough of lift coefficient are increasing, that is, the positive lift in the downstroke and the negative lift in the upstroke both increase. When the frequency is small, resistance will be generated in the upstroke at the beginning. When the frequency is greater than or equal to 33.4Hz, thrust will be generated in the whole cycle. The higher the frequency is, the greater the thrust will be. The thrust peak generated in the downstroke process is twice as much as that in the upstroke.

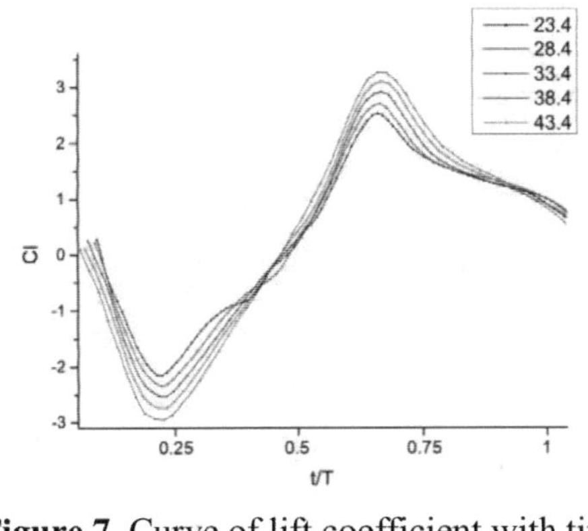

Figure 7. Curve of lift coefficient with time in a cycle

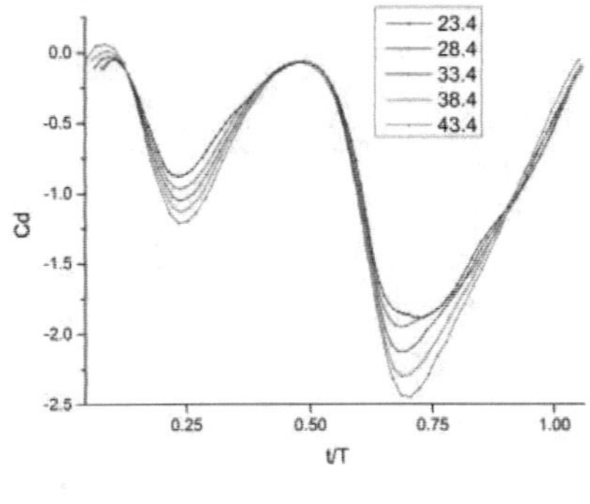

Figure8. Curve of drag coefficient with time in a cycle

4. Conclusion

In this paper, the aerodynamic performance of the flapping of the corrugated forewing of dragonfly is analyzed, and a three-dimensional model of "flapping-twisting" is established.By using the computational fluid dynamics software x-flow, the influence of flapping frequency and inflow velocity on flapping wing lift coefficient and drag coefficient was studied. Some laws of aerodynamic characteristics of corrugated forewing during flapping were obtained.

By increasing the flapping frequency and inlet velocity of the forewing, the lift coefficient can be improved. During a flapping period, the leading edge vortex (LEV) does not fall off, which is the main reason for the high lift mechanism. During the upstroke and downstroke, a thrust peak is generated at the strong trailing edge vortex, and the energy of the trailing edge vortex may provide the thrust during the forward flight.

5. Reference

[1] May M L 1991 *Dragonfly flight: power requirements at high speed and acceleration* (Journal of experimental biology) pp 325-342
[2] Reavis M A and Luttges M W 1988 *Aerodynamic Forces Produced by a Dragonfly* (AIAA Aerospace Sciences Meeting) pp 88-330.
[3] James R Usherwood and Fritz-Olaf Lehmann 2008 *Phasing of dragonfly wings can improve aerodynamic efficiency by removing swirl* (Journal of the royal society) pp 1303-1307
[4] Zhang Xingwei 2011 Computational study on the aerodynamic characteristics of the flapping wings of micro air vehicles (Harbin: Harbin Institute of Technology)
[5] Kesel A B 2000 Aerodynamic characteristics of dragonfly wing sections compared with technical aerofoils (Journal of Experimental Biology) pp 3125-3135
[6] Sun M and Lan S L 2004 A computational study of the aerodynamic forces and power requirements of dragonfly (Aeschna juncea) hovering (Journal of Experimental Biology) pp 1887-1901
[7] Wang H and Zeng L 2003 Measuring wing kinematics, flight trajectory and body attitude during forward flight and turning maneuvers in dragonflies (Journal of Experimental Biology) pp 745-757.

Acknowledgments

This work was supported in part by the National Natural Science Foundation of China under Grant 11862017, in in part by the National Natural Science Foundation of China under Grant 11462015.

IOP Conf. Series: Materials Science and Engineering **816** (2020) 012006 doi:10.1088/1757-899X/816/1/012006

An Engineering Method for Computing the Aerodynamics Performance of Hypersonic Vehicle

L Wang, G Y He, Q Wang, L S Chen

School of Aircraft Engineering, Nanchang Hangkong University, Nanchang 330063, China

HEGY509@ 163.com

Abstract: For fast and automatic calculation of aerodynamic in conceptual design of hypersonic vehicles, an aerodynamic fast calculation program based on C/C++ is compiled in this paper. In this paper, the HL-20 lift body and the double radial ball model are selected for the example verification, and the calculation results are compared with the wind tunnel experimental data, it is shown the calculation results is highly consistent with the experiment. The program not only can accurately predict the aerodynamic characteristics of hypersonic vehicles, but also be linked to the optimization program as a C++ library file to realize the shape optimization design.

1.Introduction

Near space hypersonic vehicle has the advantages of fast speed, moderate height, fast military response and strong penetration resistance. It has attracted the attention of major powers in the world and become the research frontier in aerospace field. Different from traditional supersonic and subsonic vehicles, the aerodynamic shape design of hypersonic vehicles is highly integrated. The aerodynamic shape designed during the conceptual design phase largely determines the aerodynamic performance of the hypersonic vehicle. Therefore, to obtain a suitable aerodynamic shape, large number of aerodynamic performance calculations are required, but the conventional CFD calculation method and wind experiments are inefficient in the conceptual design stage. Comprehensive considerations, engineering estimates are more suitable for this stage.

Many scholars concentrated on the aerodynamic engineering calculation of hypersonic vehicle and put forward many engineering estimation methods, such as Newton formula, Dahlem-Buck formula, tangent cone formula, expansion wave [1-4] formula and so on. Based on the above estimation formula, we can use the panel method to implement the aerodynamic estimation program for a hypersonic vehicle. In order to realize the fast calculation of aerodynamics of aerodynamic in conceptual design and optimization design of hypersonic vehicles, an aerodynamic fast calculation program based on C/C++ is compiled in this paper. The program can quickly and accurately predict the aerodynamics of hypersonic vehicles, and the results can be visualized in Tecplot.

2.Program Framework of Aerodynamic Engineering Calculation for Hypersonic Vehicle

The aerodynamic engineering program of hypersonic vehicle consists of four parts: grid reading, grid correction, aerodynamic calculation and result output.

IOP Conf. Series: Materials Science and Engineering **816** (2020) 012006 doi:10.1088/1757-899X/816/1/012006

2.1 Read the mesh model data

The STL triangle mesh generated directly by preprocessing software. In the modeling ontology coordinate system, each triangle element is represented by three vertex coordinates one normal vector, which are A(x_A, y_A, z_A),B(x_B, y_B, z_B),C(x_C, y_C, z_C) and N. Taking A as the starting point and ending with B and C respectively, two vectors can be obtained as $\boldsymbol{T_1}$ and $\boldsymbol{T_2}$.

$$\boldsymbol{N} = \boldsymbol{T_1} \times \boldsymbol{T_2} \quad (1)$$

By cross multiplication of vectors $\boldsymbol{T_1}$ and $\boldsymbol{T_2}$., the normal vector $\boldsymbol{N}$ of the plane element can be obtained. The data of each element is stored in a structure, and the structure is managed by using an alternating number tree (ADT) to realize fast storage and management of each element.

2.2 Correct the vector direction of the element

The normal vectors N calculated during the grid reading process are not all pointing to the outside of the model. But the calculation of the impact angle requires that the vector N be directed outside the model. Therefore, it is necessary to check and correct the normal vector of each plane element, which is an important prerequisite to improve the calculation accuracy. In reference [5], a fast method of modifying the vector outside the grid is presented. The ray casting algorithm is used to modify the normal vector of each plane element quickly and accurately.

2.3 Calculation of impact angle

By solving the geometric relationship of the velocity vector in the plane coordinate system, the impact angle δ between the free flow and the aircraft surface can be calculated. The calculation formula of impact angle δ is as follows:

$$\boldsymbol{\delta} = \frac{\pi}{2} - \boldsymbol{cos}^{-1} \frac{-N \cdot V}{|N|\,|V|} \quad (2)$$

The value range of δ is $-\frac{\pi}{2}$ to $\frac{\pi}{2}$.

According to the impact angle δ, the flow field of the panel element can be divided into windward (δ > 0) or leeward (δ < 0). For these two cases, different calculation formulas are used to calculate the surface pressure coefficient.

2.4 Calculation formula of pressure coefficient

In this procedure, the calculation of the windward side is the Schaaf and Chambre formula of M. C. Fadgyas [6], and the pressure coefficient of the leeward side is calculated using the ACM empirical formula [7]. The calculation formula is as follows:

2.4.1 Schaaf and Chambre formula

$$C_p = \frac{1}{s^2}\left[\begin{array}{c}\left(\frac{2-\sigma_N}{\sqrt{\pi}} s\sin(\theta) + \frac{\sigma_N}{2}\sqrt{\frac{T_w}{T_\infty}}\right)e^{-(s\sin(\theta))^2} + \\ \left\{(2-\sigma_N)\left((s\sin(\theta))^2 + \frac{1}{2}\right) + \frac{\sigma_N}{2}\sqrt{\frac{\pi T_w}{T_\infty}} s\sin(\theta)\right\}(1+erf(s\sin(\theta)))\end{array}\right] \quad (3)$$

$$\boldsymbol{C_\tau} = \frac{\sigma_T \cos\theta}{s\sqrt{\pi}}\left[e^{-(s\sin\theta)^2} + \sqrt{\pi} s\sin\theta(s\sin\theta)\right] \quad (4)$$

$$\mathbf{s} = \frac{V_\infty}{\sqrt{2RT_\infty}} \quad (5)$$

where $\boldsymbol{\sigma_N}$ and $\boldsymbol{\sigma_T}$ are normal and tangential momentum accommodation coefficients. Specular reflection simplifies to $\boldsymbol{\sigma_N} = \boldsymbol{\sigma_T} = 0$ while the diffusive reflection $\boldsymbol{\sigma_N} = \boldsymbol{\sigma_T} = 1$.

2.4.2 ACM empirical formula

$$\boldsymbol{C_p} = \boldsymbol{max}[-(\delta/15)(1/Ma^2),\ -(1/Ma^2)] \quad (6)$$

The pressure coefficient of each panel is calculated by the above equation, and the $\boldsymbol{C_p}$of the calculation result is stored in each structure.

IOP Conf. Series: Materials Science and Engineering **816** (2020) 012006 doi:10.1088/1757-899X/816/1/012006

2.5 Aerodynamic coefficient calculation and result output

According to the impact angle δ and free flow conditions, the aerodynamic parameters can be calculated:

$$C_L = -\sum C_P n_X \Delta A / S_{ref} \sin\alpha + C_z \cos\alpha \tag{7}$$

$$C_D = \sum C_P n_X \Delta A / S_{ref} \cos\alpha + \sum C_P n_y \Delta A / S_{ref} \sin\beta + \sum C_P n_z \Delta A / S_{ref} \sin\alpha \cos\beta \tag{8}$$

$$C_m = \left[\sum C_p (x - x_{CG}) n_z \Delta A - \sum C_p (x - x_{CG}) n_z \Delta A\right] / \left(S_{ref} b_{ref}\right) \tag{9}$$

Where b_{ref} is the characteristic length of the aircraft, x_{CG} is the position of the center of gravity, ΔA is the element area and S_{ref} is the reference area.

The output of the calculation result is divided into two parts: the output of the aerodynamic calculation result and the DAT file output for post processing.

3. Program Verification

In order to verify the accuracy of the calculation program, the following two models are selected to verify the calculation program: the high lift body (HL20) model designed by NASA Langley center [7] and the double ellipsoid model [8] for hypersonic atmospheric reentry。

3.1 Verification of HL-20 example

The HL-20 lift body model is derived from the laser-based HL-20 model provided by NASA. The geometric model file output by OpenVSP is shown in Figure 1. The model is imported into the pre-processing software to generate the STL grid file as shown in Figure 2:

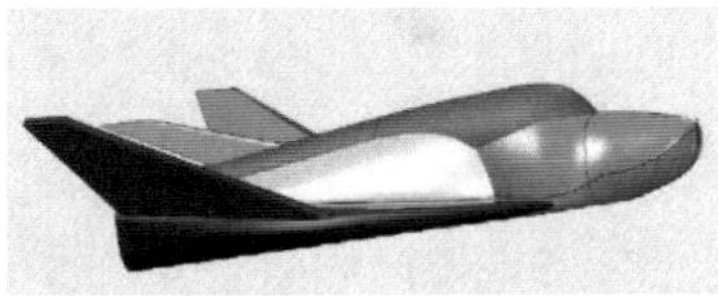

Figure .1 HL-20 lift body model

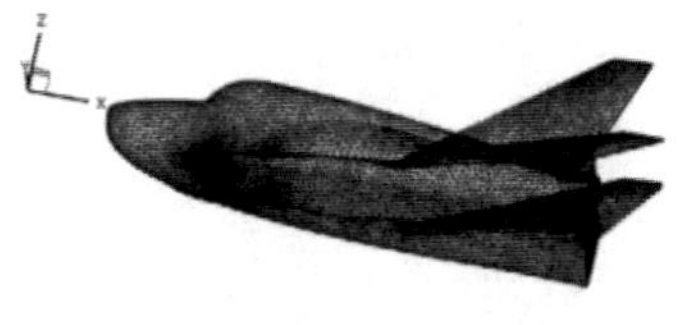

Figure .2 HL-20 lift body mesh model

In this paper, we choose the experiment of Mach 6 to verify the program. The Surface pressure coefficient distribution map of HL-20 is shown in Figure 3. And the calculation results are compared with the wind tunnel test data, as shown in Figure 4.

Figure 3. Surface pressure coefficient distribution map of HL-20

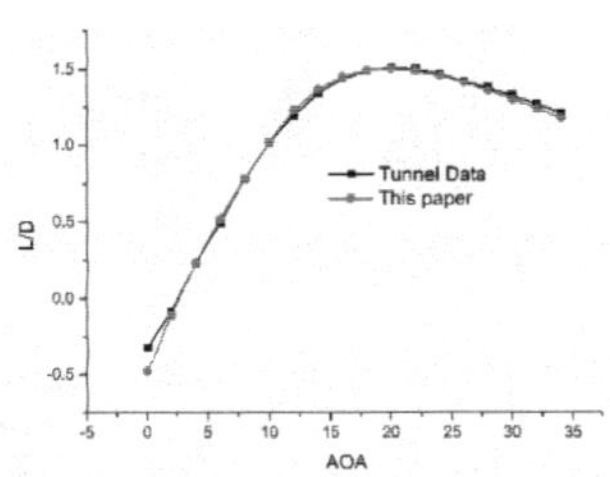

Figure 4. Comparison between calculation results and experiments

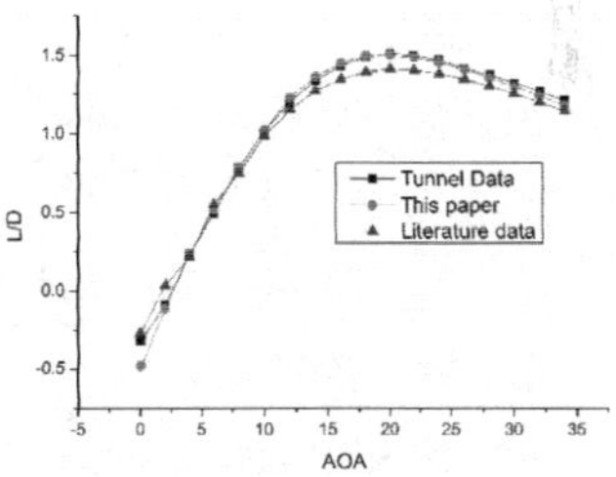

Figure 5. Comparison with Li's calculation results

Figure 3 shows the surface pressure coefficient distribution of HL-20 lift body model at 30 ° angle of attack, which is consistent with the pressure distribution at high angle of attack. It can be seen from the figure that the results obtained by the calculation program are basically consistent with the basic change trend of the experimental data of the wind tunnel, and the values are also basically consistent, with the maximum error within 10%. The accuracy of the program is verified, which can meet the accuracy requirements of the preliminary aerodynamic shape optimization design stage of the hypersonic vehicle. Compare the results of the modified Newton formula and the Prand-Meyer expansion wave formula under the same conditions in the literature[5], as shown in Figure 5. The comparison shows that the

IOP Conf. Series: Materials Science and Engineering **816** (2020) 012006 doi:10.1088/1757-899X/816/1/012006

calculation results in this paper are in good agreement with the wind tunnel experimental data. This also shows that the evolving estimation method provides an improvement in the accuracy of the aerodynamic estimation of hypersonic vehicles, which greatly improves the reliability of the calculation results.

3.2 Verification of Double ellipsoid model example

The double ellipsoid case is one of the standard cases to study hypersonic vehicle, which has a lot of wind tunnel experimental data. Its geometric model can be accurately constructed by formula. We choose this model to verify the accuracy of the calculation of the surface pressure coefficient Cp. The geometry diagram of the double ellipsoid model [8] is shown in Figure 6， The modelling of the completed double ellipsoid model is shown in Figure 7:

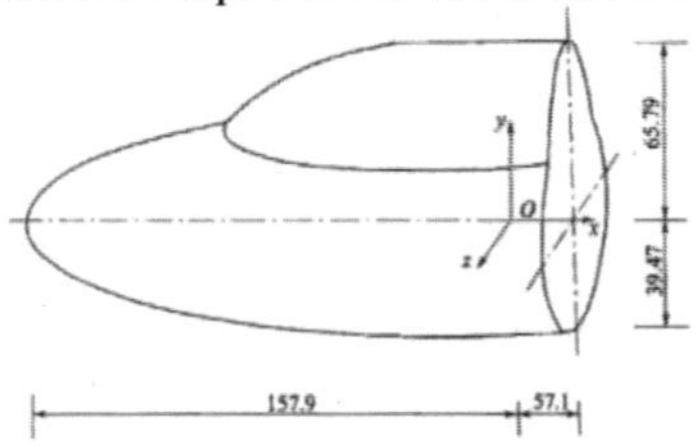

Figure 6. Double ellipsoid model

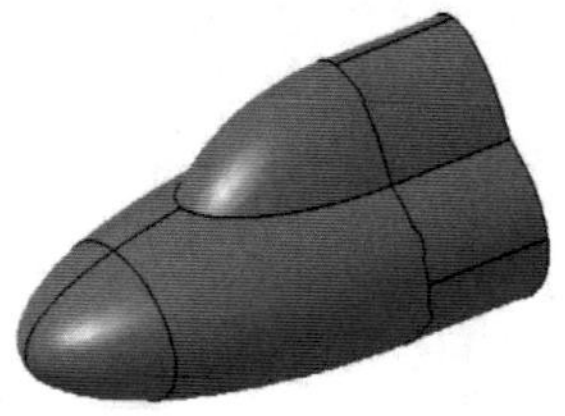

Figure 7. Double ellipsoid model

The shape of the double ellipsoid model can be modelled by the following formula:

Horizontal ellipsoid $$\left(\frac{x}{157.9}\right)^2+\left(\frac{y}{39.47}\right)^2+\left(\frac{z}{65.79}\right)^2=1 \quad (10)$$

Vertical ellipsoid $$\left(\frac{x}{92.11}\right)^2+\left(\frac{y}{65.79}\right)^2+\left(\frac{z}{46.05}\right)^2=1 \quad (11)$$

Upper half cylinder $$\left(\frac{y}{39.47}\right)^2+\left(\frac{z}{65.79}\right)^2=1 \quad (12)$$

Lower half cylinder $$\left(\frac{y}{65.79}\right)^2+\left(\frac{z}{46.05}\right)^2=1 \quad (13)$$

In this paper, we select the experimental data of the surface pressure distribution of the double ellipsoid (cannon wind tunnel) for comparison. The free-flow Mach number is 7.79 Ma, the Reynolds number per unit length is 1.04×10^7/m, the angle between the airflow and the long axis of the double ellipsoid is 0°. Compare the distribution of surface pressure coefficients on the center line. The comparison of wind tunnel test data of calculation results is shown in Figure 8 and Figure 9:

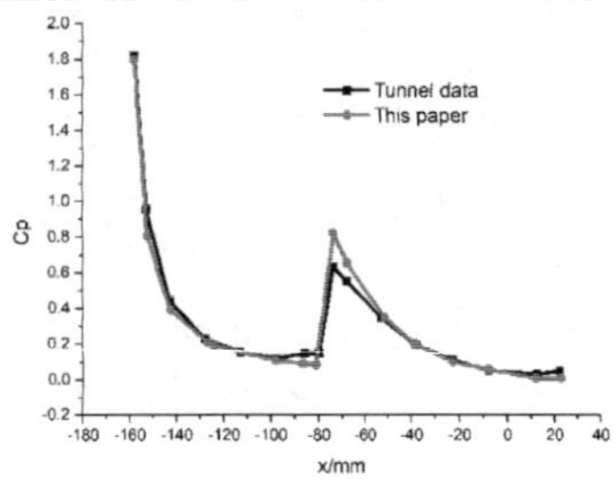

Figure 8. comparison of partial pressure coefficient of the upper surface of the center line

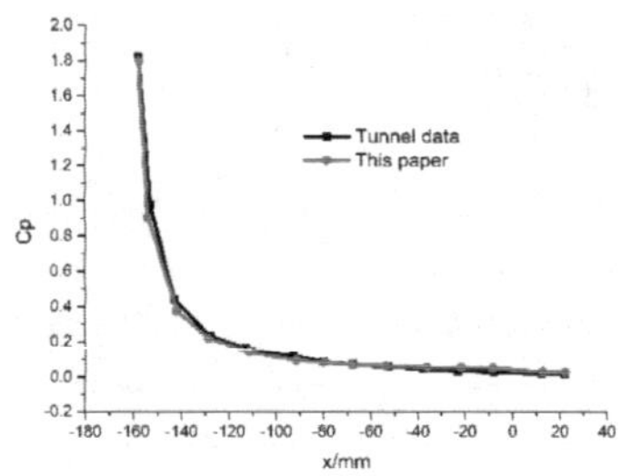

Figure 9. comparison of partial pressure coefficient of the lower surface of the center line

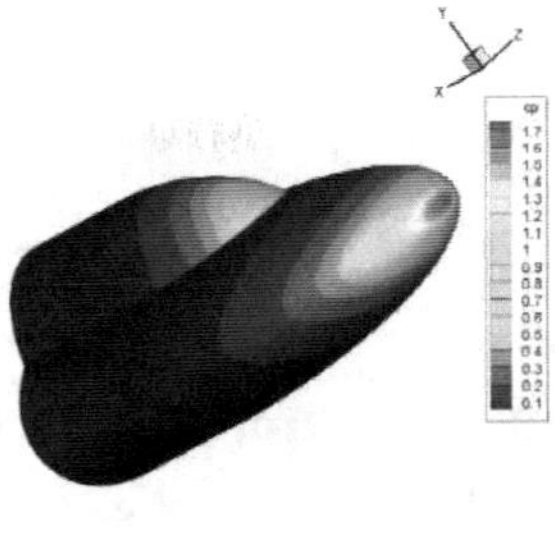

Figure 10. Surface pressure coefficient of the angle of attack 0° double ellipsoid model

Figures 8 and 9 show a comparison between the wind tunnel experimental data and the calculation results of the program. Figure 10 shows the surface pressure coefficient distribution of the double ellipsoid model at an angle of attack of 0°. The calculation data in Figs. 8 and 9 is extracted by the post-processing software in the program-generated DAT file. It can be seen from the comparison chart that the calculation results of the pressure coefficient of the lower surface are in good agreement with the experiment. On the upper surface, the calculation results are basically consistent with the experimental data, and the numerical values are basically the same, but there are some deviations in the -80mm to -50mm part. Comparing Figure 8 and Figure 9, the range of -80mm to -50mm belongs to the intersection of the upper ellipsoid and the lower ellipsoid. In this range, the pressure coefficient changes greatly with the change of the x position, and there are some differences between the position of the experiment and the position of the calculated result, which leads to the deviations. In general, the accuracy of the calculation of the surface pressure coefficient Cp fully meets the engineering calculation requirements.

4. Conclusion

Two examples of HL-20 and double ellipsoid model are calculated, and the results are compared with the experimental data of wind tunnel. The following conclusions can be drawn:

1. The program designed in this paper can quickly and accurately estimate the aerodynamic force of hypersonic vehicle. And the results can be visualized in the post-processing software Tecplot.
2. Compared with the experimental data and other literature results, the estimation formula used in this paper (the combination of Schaaf and Chambre formula and ACM empirical formula) has higher calculation accuracy, which can fully meet the accuracy requirements of engineering calculation.

References

[1] Anderson Jr and J D 2006 *Hypersonic and High-Temperature Gas Dynamics, Second Edition* (AIAA) pp 449~457
[2] Rosema C, Doyle J and Blake W B 2014 Missile Data Compendium (Datcom) User Manual 2014 Revision
[3] Bonner E, Clever W and Dunn K 1991 Aerodynamic Preliminary Analysis System II. Part I-Theory. Contractor Report, 182076.
[4] Kinney D, Garcia J and Huynh L 2006 Predicted convective and radiative aerothermodynamic environments for various reentry vehicles using CBAERO. In 44th AIAA Aerospace Sciences Meeting and Exhibit p 659
[5] Li Zhengzhou 2017 Multidisciplinary Design Optimization for Winged Re-entry Vehicles Considering Stability and Control Characteristics. Nanjing University of Aeronautics and Astronautics pp 46-48.
[6] M. C. Fadgyas, M. V. Pricop.Semi-empirical relations for pressure distributions in hypersonic regime. International Conference of Numerical Analysis and Applied Mathematics (ICNAAM 2017), 2018：370007-1–370007-5.
[7] Moore M and J Williams 1989 Aerodynamic prediction rationale for analyses of hypersonic configurations. 27th Aerospace sciences meeting p 525
[8] Li Sunxun 2007 Typical shape hypersonic flow characteristics. National Defence Industry Press pp 18-27.

IOP Conf. Series: Materials Science and Engineering **816** (2020) 012007 doi:10.1088/1757-899X/816/1/012007

Based on Bionic Optimization Design and Strength Analysis of The Tie Rod of Aircraft Landing Gear

M T Tang[1], **Q Wang**[1*], **G Y He**[1] **and R P Liu**[1]

[1] Nanchang Hangkong University, Nanchang 330063, China

645172379@qq.com

Abstract. Based on the three-dimensional model of the cross-type tie rod in the landing gear of an aircraft, the mechanical analysis and solution are carried out, and the stress nephogram as well as the maximum position are obtained. Based on the analysis of bamboo microstructure, two kinds of imitated bamboo tie rod has been designed, one is the imitated bamboo without knots, the other is added knots. The three-dimensional modeling is carried out. The static analysis of finite element is carried out under the same working condition, and the stress nephogram and the position of the maximum value are obtained.The results of two kinds of simulated bamboo tie rod and cross-type tie rod are compared and analyzed. The results show that the mechanical properties of the whole simulated bamboo tie rod are superior to the cross-type tie rod.

1. Introduction

As one of the most important means of transportation, aircraft plays an irreplaceable role in human production and life, and landing gear is an indispensable structural part of aircraft. In the process of taking off and landing, the landing gear is a very important force bearing part. In addition, the gliding and turning on the ground are also completed by the landing gear.At the same time, with the improvement of aircraft performance, in addition to the more stringent requirements on the structural stress, strength and service life of the aircraft landing gear, the higher requirements on the weight of the aircraft landing gear are also put forward.

Aiming at the research of landing gear optimization and lightweight, the landing gear of a UAV is optimized based on the structural topology optimization method according to Yang et al[1] , and make the structure achieve better results. For the problem of strength analysis of landing gear, Yuzhen Chen et al[2]. Studied the static strength simulation analysis of landing gear axle;Xueyu He et al[3]. Carried out the static structural optimization of landing gear strut.Because the brake rod in the landing gear is the key stress component when the aircraft brakes on the ground, its strength and structure will also have a key impact on the landing of the aircraft. Therefore, the paper studies and analyzes the existing cross-type tie rod.By observing the microstructure of bamboo, a new type of brake rod with bamboo like microstructure is proposed and designed, and its strength is analyzed. It is compared with the cross-type tie rod.

2. Establishment and Static Analysis of Cross-type Tie Rod

2.1. Cross-type tie rod and its establishment of three-dimensional model

As the key part of the brake after landing, the tie rod is mainly affected by the tension, with slight bending moment and torque.At present, the shape of the brake tie rod in use is a symmetrical shape model with the cross section of the middle cylinder with ears on both sides[4].Through the analysis, it can be seen that the contact position of the tie rod in the work at the ear hole.In order to make the analysis result reflect the real situation, the bolt is installed in the ear hole of the tie rod,in order to apply loads and boundary conditions,CATIA 3D model as presented in figure 1.

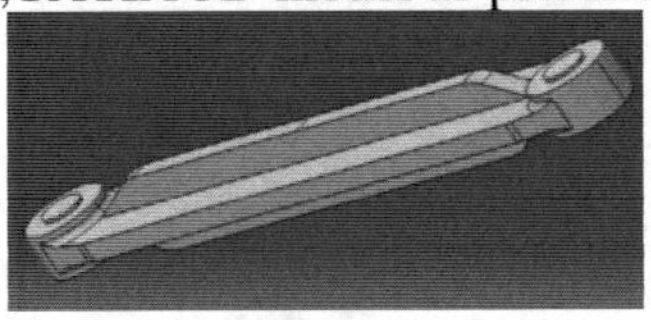

Figure 1. three-dimensional model of cross-type tie rod.

2.2. Static analysis

Firstly, the established three-dimensional model is imported into the ANSYS Workbench to generate the finite element model, and then 10 node tetrahedral element is used for automatic mesh generation. The mesh model as presented in figure 2.

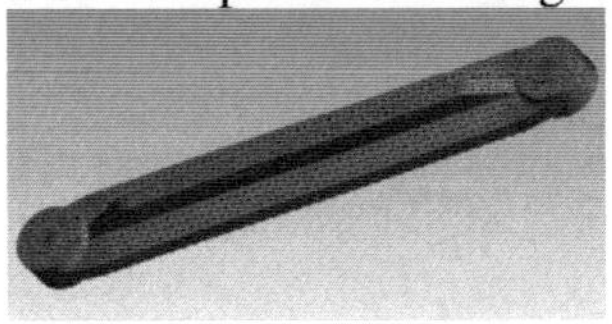

Figure 2. finite element mesh model of cross tie rod.

Figure 3. Two methods of force application for moment on tie rod.

According to the imported model coordinates and the stress analysis of the tie rod, it is concluded that the three coordinates of the bolt at one end of the tie rod should be constrained.The tension applied by the bolt at the other end is 300KN, the bending moment is 8KN, and the force and torque applied at the bolt is 2200N·m.Due to the different height of the ribbed plate in the middle of the tie rod, there are two directions when the bending moment is loaded. As seen in figure 3, the load is perpendicular to the ear bolt axis ,as case B,and parallel to the ear bolt axis,as case A.The material property[5] is 30CrMnSiNi2A high strength alloy steel, the modulus of elasticity is 211 GPa, the Poisson's ratio is 0.3, and the density is 7850 Kg/m^3.

2.3. Analysis results

According to the statics analysis and solution of the cross-type tie rod after loading under the working conditions, it is concluded that the maximum stress of the tie rod under both loading conditions occurs at the joint between the outer side of the cylindrical ear and the rib plate.As seen in figure 4, the maximum stress under the case A is dangerous.

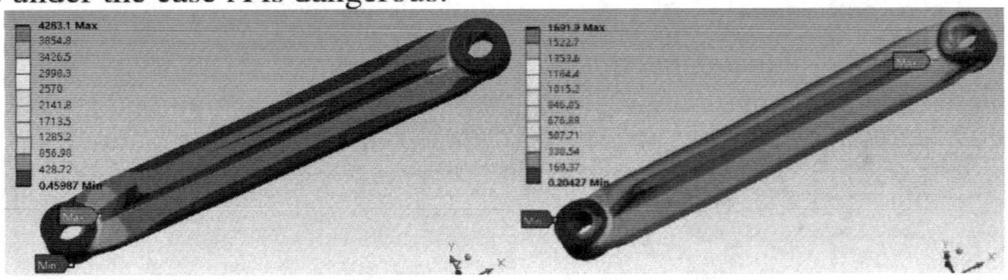

Solution results of stress cloud(The case A is left and the case B is right)

Figure 4. Solution results in two cases

3. Design and Static Analysis of Bionic Tie Rod

3.1.Microstructure analysis of bamboo

Bamboo skin is the material basis of bamboo's excellent mechanical properties, in which there are many kinds of cells with different morphological structures. From the mechanical point of view, these cells can be divided into two categories: one is the basic tissue cells, which are the thin-walled structure that plays the role of load transfer; the other is the thick walled cells mainly vascular bundles, which are surrounded by the basic tissue cells and play a bearing role.

At the same time, there are a large number of thicker tubular cells in the vascular bundle, which greatly enhance the mechanical properties of bamboo. As presented in figure 5, the density of this kind of cells in the outer layer is the highest, and then gradually becomes low to near the endothelial layer and slightly increases[6-8].

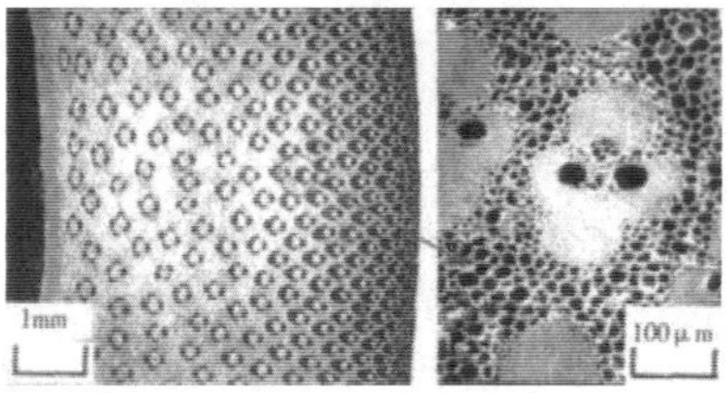

Figure 5. Microstructure of bamboo cross section

3.2 Design and modeling of bionic tie rod

According to the research and analysis of bamboo microstructure and mechanical properties, the tie rod was redesigned.In order to improve the mechanical properties, the thick walled cells of vascular bundle with load-bearing function were transformed into stiffeners. The stiffener and the wall thickness are connected to realize the function of load transfer of basic tissue cells.On this basis, imitated bamboo tie rod without knots(see figure 6) and the imitated bamboo tie rod with knots(see figure 7) are designed.

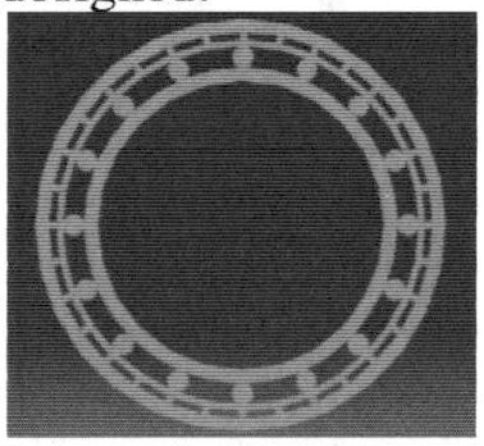

Figure 6. Cross section of imitated bamboo tie rod without knots

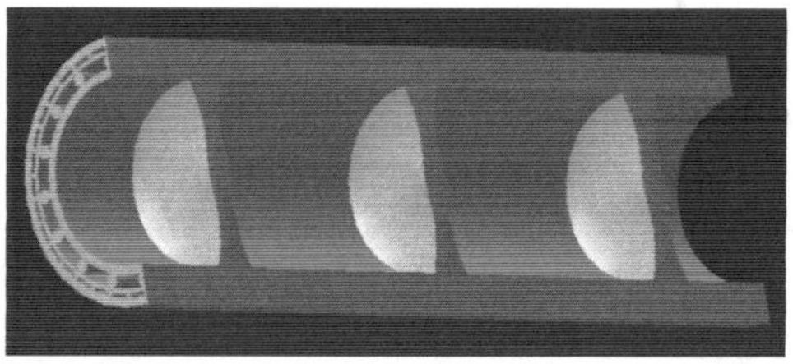

Figure 7. A model of imitated bamboo tie rod with knots

3.3 Static analysis of bionic tie rod

The redesigned two kinds of imitated bamboo tie rod models are introduced into the ANSYS Workbench,and the finite element analysis and solution are carried out according to the same mesh division mode, boundary constraints and loads as the cross-type tie rod,The solution results are shown in figure 8 and figure 9.Among them,in the case A,two kinds of bamboo imitation rods have stress concentration at the inner edge of the ear hole. However,this place is not a dangerous area by analyzing, the dangerous area is located at the hollow end of the interior to meet the actual situation, as shown in the figure. Top right picture of 8 and 9.

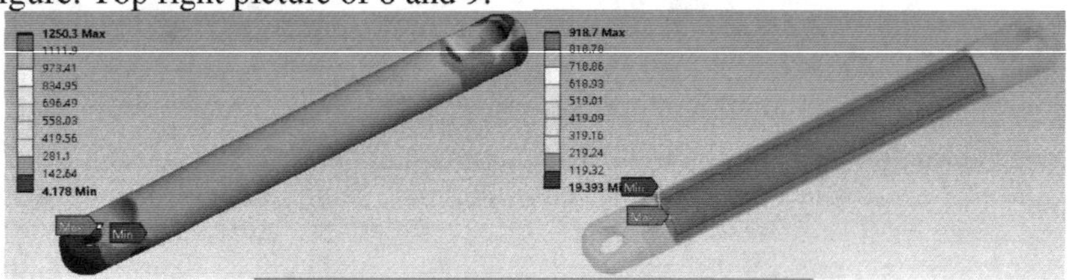

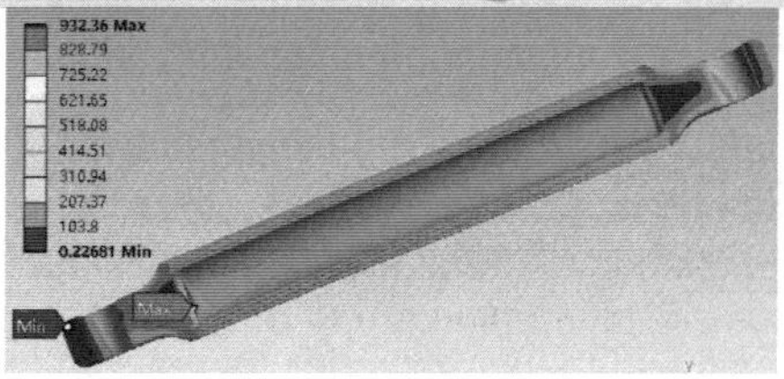

Figure 8. Stress cloud of imitated bamboo tie rod without knots in two case(The case A is above and the case B is below)

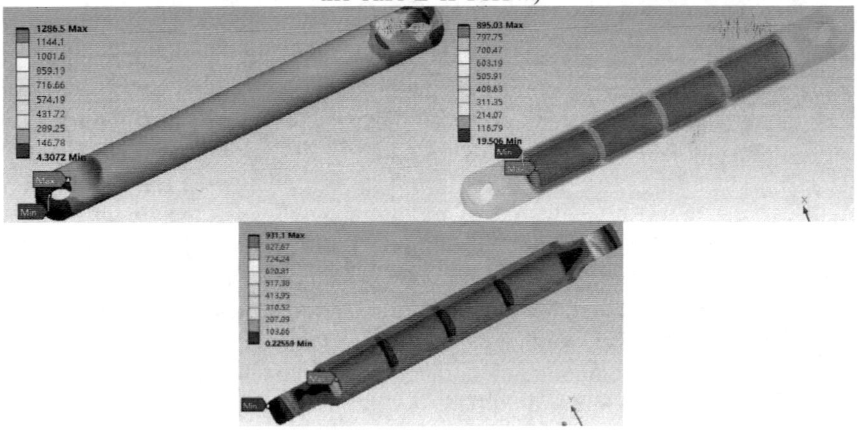

Figure 9. Stress cloud of imitated bamboo tie rod with knots in two case (The case A is above and the case B is below)

4. Comparative Analysis of Results

According to the static analysis of the cross-type tie rod, the imitated bamboo tie rod with bamboo joint and the imitated bamboo tie rod without bamboo joint, the stress nephogram is obtained, as presented in figure 10. For the convenience of comparing the results, the data summary of the corresponding stress nephogram is shown in table 1.

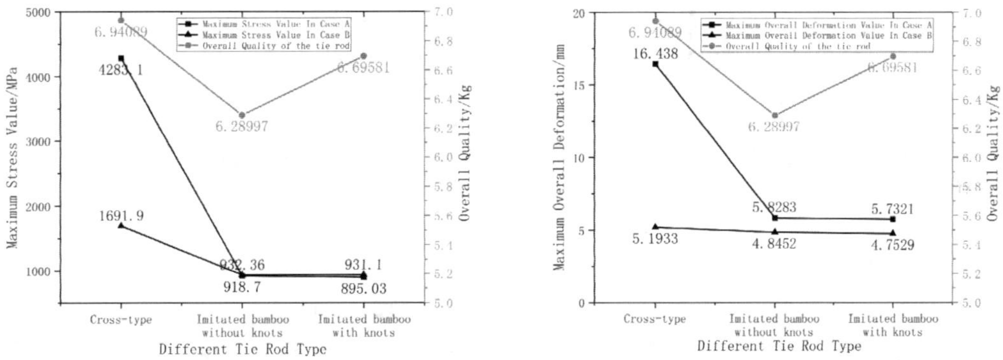

Figure 10. Data comparison line chart (Stress Value in the left and Overall deformation Value in the right)

Table 1. Comparison of statics results between cross-type tie rod and imitated bamboo tie rod.

	Cross-type tie rod	Imitated bamboo tie rod without knots	Imitated bamboo tie rod with knots
Maximum stress value in case A/MPa	4283.1	918.7	895.03
Maximum Overall deformation in case A/mm	16.438	5.8283	5.7321
Maximum stress value in case B/MPa	1691.9	932.36	931.1
Maximum Overall deformation in case B/mm	5.1933	4.8452	4.7529
Overall Quality/Kg	6.9408915	6.2899695	6.6958145

From table 1, The structural quality of the two imitated bamboo tie rods is slightly lower than that of the cross-type rods, which achieves a certain degree of lightweight structure.In the case A, the maximum stress value and the maximum overall deformation value of the imitating bamboo tie rod have greatly decreased. Both the strength and stiffness of the imitating bamboo tie rod have significantly improved. Specifically, the imitating bamboo tie rod without knots has a maximum stress value of 78.55% lower than the cross-type tie rod in the dangerous area, and the overall resistance to deformation is increased by 64.54%. The imitating bamboo tie rod with knots has a maximum stress value of 79.10% lower than the cross-type tie rod in the dangerous area,and the overall resistance to deformation is increased by 65.13%. In the case B, the maximum stress value and the maximum overall deformation value of the imitating bamboo tie rods have also decreased significantly. Specifically, the imitating bamboo tie rod without knots has a maximum stress value of 44.89% lower than the cross-type tie rod in the dangerous area, and the overall resistance to deformation is increased by 6.70%. The imitating bamboo tie rod with knots has a maximum stress value of 44.97% lower than the cross-type tie rod in the dangerous area,and the overall resistance to deformation is increased by 8.48%. Compared with the imitating bamboo tie rod without knots, the imitating bamboo tie rod with knots was slightly improved, but the quality increased by 6.06%.

5. Conclusions

In this paper, the three-dimensional model of the cross-type tie rod in the landing gear brake system of an aircraft is established by using CATIA software. The static analysis is carried out in the ANSYS Workbench finite element analysis software, and the stress nephogram of the cross-type tie rod is obtained.According to the relationship between the excellent mechanical properties and microstructure of bamboo, the cross-type tie rod was redesigned，The imitated bamboo tie rods with knots and without knots are obtained, and the comparison between them is made. At the same time, their stress nephogram and the position of the maximum value are obtained.

Through the comparative analysis of the cross-type tie rod and two kinds of imitated bamboo tie rods under the same working condition, it is concluded that both the strength and the rigidity are greatly improved under the working condition of the case A and the case B.In general, the mechanical properties of the two kinds of imitated bamboo tie rods are superior to those of the cross-type tie rods. The redesigned bionic tie rod has a high reference value for improving the quality and mechanical properties of the cross-type tie rod in the landing gear brake system of the aircraft.

6. References

[1] LIU Wenbin,ZHANG Ming,CHEN Yuhong.*Topology optimization of an UAV landing gear structure*[J].Mechanical Science and Technology for Aerospace Engineering,2014,33(11):1753-1757

[2] CHEN Yuzhen,YU Weijian.*Simulation Analysis of Aircraft Main Landing Gear Axle*[J].Computer Application Technology,2006,33(11)

[3] HE Yuehong,ZHANG Yingjie,XIE Liyang et al.*Finite element numerical analysis of dangerous parts of nose landing gear of a certain type of aircraft*[J].Structural strength analysis

[4] YAO Guangsheng.*The key strength of aircraft landing gear components analysis and structural optimization*[D]. Taiyuan:North University of China,2014

[5] JIANG Chengjie.*Research on fatigue life analysis and optimization of landing gear for a largeamphibian aircraft*[D]. Nanjing:Nanjing University of Aeronautics and Astronautics,2014

[6] MA Jianfeng,CHEN Wuyi,ZHAO Ling.*Bionic design of columnar structure based on bamboo microstructure*[J].Mechanical Design,2009,25(12):50-53

[7] HUANG Shengxia.*Relationship between the structure of bamboo and mechanical behavior*[D]. Hefei:Master's thesis of Wood Science and Technology,Anhui Agricultural University,2007:6-8.

[8] SHANG Xinlong,MAO Tengfei,GUAN Xin,et al.*Study on the Distribution Law of Bamboo Fiber in Natural Column*[J].FRP/Composite material,2013(3):93-96.

IOP Conf. Series: Materials Science and Engineering **816** (2020) 012008 doi:10.1088/1757-899X/816/1/012008

Wind Tunnel Testing Verification Method for Flight Test Pilots Emergency Departure from Commercial Aircraft

Yang Liu

Shanghai Aircraft Design and Research Institute, Shanghai, 201210, China

auvpomelo@126.com

Abstract. According to airworthiness regulations, the first or previous several commercial aircraft for flight test must be installed with an air emergency departure system to ensure that the members of the pilots safely escape from the aircraft in the event of an irretrievable flight emergency. Large-scale commercial aircraft mostly adopt the lower-single-wing and wing-hanging-engine configuration. Emergency departure channel is generally designed on the side or bottom of the front fuselage, when pilots had to leave the aircraft, who may hit the aircraft components, such as landing gear, engines and wings. Therefore, it must be verified whether the test pilot can safely leave the aircraft from the emergency exit during the aircraft design process. This paper introduces the method about how to verify the safety of test pilots though wind tunnel testing. The key of method is to measure the trajectory of pilot in wind tunnel by simulating conditions of flight and pilot emergency departure. The method illustrated can be used for design and verification of emergency departure system.

1. Introduction

The emergency departure system is an indispensable system for the first or previous several commercial aircraft used for flight test according to airworthiness regulations [1]. Mmedium and long range commercial aircrafts, including Bombardier's C Series, Airbus A350 and A380, are equipped with such system. Normally the structure, avionics and other related systems will be modified to meet the airworthiness requirements for aerial emergency departure based on the prototype [2].

The key to verify the safety of the air emergency departure is to obtain the trajectory of the test pilot from emergency departure exit. Generally, numerical calculation, wind tunnel testing and flight test could be used to verify safety of flight test pilots emergency departure. Although the numerical calculation has the advantages of short cycle and low cost, it is difficult to meet the requirements of the accuracy of dynamic flow field simulation under complex flight conditions[3], especially in the stall state. Wind tunnel testing can not only simulate complex flow fields, but also has high accuracy to simulate the test pilot leaving the aircraft. The flight test is used through projecting the dummy pilot model from the exit to judge the safety of the trajectory, and this method is relatively risky, which cost a lot and lags behind design and development. In summary, the wind tunnel test is the most suitable and reliable verification method.

2. Testing Principles

In order to ensure the trajectory of the center of gravity of the model and the attitude of turning around the center of gravity is similar to the real object, the following principles should be met which are that the model's geometric shape should be similar to the real object, the air speed for wind tunnel testing and real flight are equal and dynamical similarity between the model and the real object. In other words, not only the aerodynamic force acting on the model, but also the effect of the model's gravity [4] must be considered.

The wind tunnel testing should follow dynamical similarity. The testing speed is generally low, therefore the effect of Mach number is not considered and the effect of Reynolds number is negligible. Froude number should be Equal, that is:

$$\frac{V2_m}{l_m g_m} = \frac{V2_f}{l_f g_f} \tag{1}$$

The subscript *m* in formula (1) represents state of the scaled model, and *f* represents state of the full-scale model. *V* represents speed, *l* represents characteristic length, *g* represents gravitational acceleration and *m* represents mass

According to the dynamic similarity rule of formula (1), the following similarity conditions can be obtained:

a) Speed similarity criterion:

$$u_m = u_f/\sqrt{K} \tag{2}$$

In formula (2), *u* represents speed, and *K* represents the inverse of the model scale ratio, $K = l_f/l_m$

b) Quality similarity criteria:

$$m_m = \frac{m_f}{K^3 \Delta} \tag{3}$$

$\triangle$ represents the relative density of air, $\Delta = \rho_f/\rho_m$

c) Similarity criterion of moment of inertia around the center of mass:

$$I_m = \frac{I_f}{K^5 \Delta} \tag{4}$$

In the formula (4), *I* represents the moment of inertia around the center of mass.

3. Testing Equipment and Model

3.1. Wind tunnel and measuring equipment

As to the wind tunnel selection, the bigger the size of wind tunnel section is, the better the similarity is. Normally it is better to choose the wind tunnel with a section larger than 3m×3m, which will make the pilot easy to keep similarity. Generally, low-speed force&moment measuring wind tunnels larger than order of 3 meters' section can meet the requirements of tests, such as DNW LLF in the Netherlands, CARDC FL-12 \ FL-13 and AVIC FL-9/FL-10 in China, etc. The emergency departure test is different from force&pressure test, and delivery of test is trajectory data of the pilot model. Because emergency departure is an instantaneous dynamic process, and the duration usually lasts about 0.2s-0.3s, high-speed (HD) camera should be equipped to capture the trajectory. HD cameras need to be installed in three positions at least, which are better to be located at the entrance of the test section, the side of the test section and the bottom of the test section.

3.2. Aircraft model

The full model of aircraft could fulfill the requirements of different positions of emergency exit, and the components, such as fuselage, wings, engines, tails, landing gears, should be included.In addition, the model could realize deflection of slat &flap and retraction of landing gear to simulate various flight conditions of the aircraft.The model scale is determined according to the requirements of the wind tunnel blockage. The requirements of model design and manufacture are consistent with the requirements of the conventional wind tunnel model.

3.3. Model support system

The emergency departure exit of commercial passenger aircraft is generally located at the side or bottom of the front fuselage, and the trajectory of the pilot is below the fuselage and wings. Therefore the model should be supported with rear sting or back sting to reduce the affluence of support sting.

3.4. Pilot model
The pilot model includes human body model and parachute carried. The similarity of the pilot model with pilot should be considered, including external dimensions, weight, posture, and moment of inertia around the center of gravity should be. The overall dimensions and weight of pilot and parachute can be formulated according to standard. The posture of the human body is calculated based on the statistics of the pilot's parachuting posture. The general posture is mostly the two-handed knee-guard posture. The pilot model could be designed in CATIA, and the moment of inertia of the model around the center of gravity can also be calculated from the CATIA. The size, weight, center of gravity, and moment of inertia of the model will have a significant influence on the test results, therefore, the model should be fabricated strictly in accordance with the 3D geometry of the model and debugged according to the requirements of mass and moment of inertia. The detection review should be performed to ensure the model can reflect the true situation of pilot. The pilot model may be damaged in the wind tunnel, so a certain number of backup models need to be prepared.

3.5. The departure device
The departure device is used to ensure that the test pilot model could leave the aircraft smoothly from the departure channel during the wind tunnel testing, which should be designed according to the location of the exit and the way of departure. Two types of devices could be used, which are free departure devices and pressured departure devices. The free departure device could ensure the model departure by cutting off the connection between the pilot model and the aircraft model. Normally the pilot model could be connected to the aircraft model by fusible constantan wire or magnetic connection. Preassured departure devices realizes the departure of the pilot model by acting pressure on pilot model, and the pressure is provided with high pressure gas source from 1 to 15Mpa.

4. Testing Conditions and Arrangements

4.1. Testing conditions.
Emergency departure of pilots will take place with irreparable situation during the flight test. Irreparable flight test conditions, such as stall, will be critically simulated in wind tunnel. Theses typical test conditions include terminal conditions of altitude, stall speed, aircraft attitude, angle of attack at stall and the takeoff and landing configuration. Stalling is used as a typical example to illustrate selection of the dangerous working conditions.

4.1.1. Altitude. Air density vary slightly in low altitude from 200m to 1000m, when stalling occurs , it is very difficult to recover in the case of high descent rate, so the critical altitude allowing the parachute open is about 200m. Because there are more than one pilot during the test flight, 700m is selected as the test altitude for low altitude in order to ensure that the test pilots has time to jump off the aircraft. 3000m and 5000m are as middle altitude.

*4.1.2. Speed. S*tall speed and the maximum skydiving speed are considered as limited speed. Maximum skydiving speed is determined according to the type of door at the exit of the aircraft [5], and the stall speed of takeoff, cruise, and landing are selected at the maximum weight.

4.1.3. Attitude angle. Flight attitude is also an important condition that affects the trajectory, so the angle of attack at stall and a relatively large side slip angle (± 10 °) are selected for wind tunnel testing. In addition, the initial departure speed of the pilot model and the anti-air blow device are also important conditions that affect the trajectory of the aircraft.

4.2. Testing content

Wind tunnel testing could be carried out by method of controlling variable, and the variables include conditions of altitude, speed and attitude angle for each model configuration. Each run of test monotonously changes one variable A and keeps the other variables constant. Then, monotonically change another variable B and keep the other variables constant. Due to the large number of variables, the runs of test will be huge if the test is performed exactly as described above. In order to improve this situation, the terminal envelope method can be used to simplify test runs, that is, to perform tests under various terminal test conditions to obtain the test results in extreme cases, and then derive other conditions according to the law.

5. Testing Procedure and Results Analysis

5.1. Testing steps

The testing procedure is different from the conventional force&pressure test, which mainly includes the following steps.

a) Mount aircraft model in the wind tunnel test section with back support or rear support, and install the HD cameras. Run the test at the given speed and model attitude, and start the departure device to make the pilot model separate from the aircraft model under the given initial speed, the pilot model will keep rolling down with the action of gravity and aerodynamics.

b) Capture the motion trajectory and attitude of the pilot model and check whether the pilot model is leaving from exit with expected initial speed and attitude, then record the data if the conditions are assured.

C) Inspect whether the pilot model is damaged after the each run, including inspection of the moment of inertia. If model is damaged, it should be repaired as much as possible. If the model cannot be repaired, replace the model with backup.

d) Each run should be interpreted and analysed, and then the test parameters are adjusted according to actual needs.

5.2. Results analysis

According to HD video, the motion trajectory of the pilot model at the boundary conditions could be drawn, and dangerous boundary could be recognized according to the distance between the pilot model and the aircraft component. Meanwhile impacts of the altitude, flight speed, angle of attack, side slip angle, and different configuration could be given. The results could be divided into three levels, which are safety, danger and collision. The safety means that motion trajectory of the pilot model is far away from the aircraft model. Danger means that motion trajectory of the pilot is close to components (such as the engine, wings, landing gear, etc.) without colliding with the aircraft model. Collision means that the pilot model collided with the aircraft model.

6. Conclusions

Wind tunnel testing verification method for emergency departure of pilots from commercial aircraft is demonstrated in this paper, including the selection of wind tunnels and equipment, pilot model design, test conditions setting, test steps, and results analysis. The method could be used to provide technical basis and reference for airworthiness verification of commercial aircraft emergency departure.

References

[1] Civil aviation administration of China 2007 CCAR-21-R3 *Airworthiness regulations of civil aviation product and components.*, Section 35, Flight Test (4).

[2] Australian Government Aviation Safety Authority 2012 AC 21-47 *Advisory Circular-Flight test safety.*

[3] Dongyun Zhang and Meili Wang 2015 *Computational Method for Emergency Departure Trajectory of Civil Aircraft* (China Science and Technology Information) V13.

[4] Jiechuan Fan 2002 *Wind tunnel test handbook*, Beijing: Aviation Industry Press.

[5] Qingyuan Sun 2004 *Flight Test Method for Emergency Escape from Military Transport Aircraft* (Flight Test), P25.

Three-Dimensional Reconstruction of Continuous Tomographic ICT Images of Defective SRM

Ze Zhang*, Feng Gao, Qian Zhang

Air and Missile Defense College, Air Force Engineering University, 710051 Shaanxi Xian, China

*Email: zhangze_xyz@163.com

Abstract. As an advanced technology for non-destructive testing of defective SRM, industrial computed tomography can visually show the structure of the SRM's transverse section. 3D reconstruction of continuous tomographic images technology can convert the 2D images obtained into 3D visualization models, thus accurately determine the size and distribution of internal defects in the SRM, etc. Study the main technical methods of 3D reconstruction, it mainly includes 2D images denoising, edge detection and gray-scale image segmentation. Project, accumulate and render the pre-processed 2D images to obtain the 3D model, and then smooth the surface of the 3D model, the end result is a clear reconstruction model. The reconstructed 3D model can provide efficacious data support for SRM non-destructive testing, making the testing results more real and effective.

1. Introduction

During the long-term storage and on-duty work of solid rocket motor (SRM), due to environment, load and other factors, the SRM will produce different degrees of defects and damage. These defects are most likely to expand under the SRM operating conditions, causing the SRM become invalid and even exploded. For a long time, due to the lack of advance detection technology, in order to ensure safety, engineers had to eliminate some defective motors but work fine, resulting in a great waste of resources. So, it is of great strategic significance to evaluate and predict the performance of defective SRM.

Industrial Computed Tomography (ICT) as an advanced technical means for non-destructive testing of SRM, the ICT images obtained from the scan can reflect the structure of the propellant grains in the slice. In medicine, doctors mainly use the medical knowledge and experience to judge the structure of slice images to achieve the purpose of diagnosis. However, this method relies on the personal experience of the doctor, for engineers, they often do not have the relevant knowledge to directly analyze the ICT images, therefore, it is difficult to make an effective analysis of the overall performance of the SRM with only 2D images. This makes the 2D images inverse 3D model become a more effective way to solve this engineering problem. This technology can visualize the internal structure of the SRM, allowing researchers to directly analyze and process the 3D model they have built [1, 2, 3].

Since the 3D reconstruction is a process from discrete to continuous, it will inevitably result in the loss of some information. The problem that arises from this is the contour mismatch between layers, so the accuracy of the reconstructed model will also be affected. The most direct way to improve the accuracy of models is to reduce the scanning interval. However, when the resolution of the existing

ICT machine is not enough to scan the slice at a smaller interval, it is particularly important to choose an appropriate reconstruction method.

2. Reconstruction method

In the current research, the method used are mainly divided into surface rendering and volume rendering. The advantage of the surface rendering method is that the model generation speed is fast, but this method only reconstructs the surface of the measured object, and cannot obtain data information inside the object. The volume rendering method can be used to display the internal structure of the object, but this method requires a large amount of information in the calculation process, so the model generation speed is low [4].

2.1. *Surface rendering*

In the process of 3D reconstruction by surface rendering method, firstly, the image is processed for noise filtering and edge detection, and then the detected edge information is subjected to inter-layer interpolation processing, thereby splicing to obtain the object surface and smoothing the surface, finally, rendering to get a 3D surface model. The main process is shown in Figure 1. Typical surface rendering algorithms mainly include Connected Contour, Cuberille, Marching Cube and Dividing Cube [6].

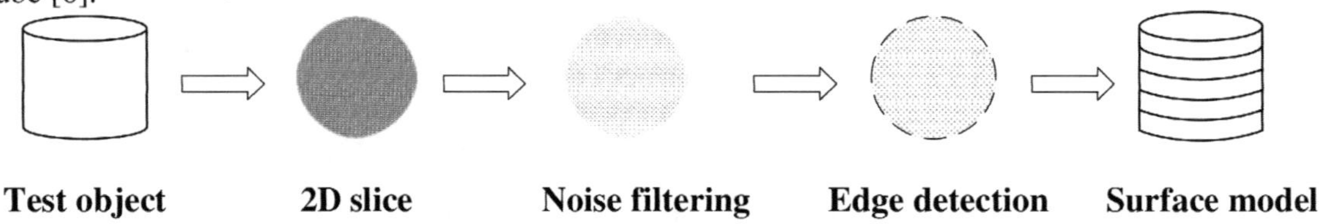

Figure 1. The main process of surface rendering method

2.2. *Volume rendering*

The central idea of volume rendering is to first analyze the transparency and brightness of the voxels, and then calculate the emission, reflection and transmission of light through the voxels to obtain a projected image. Typical volume rendering algorithms include Ray-Casting, Splatting and Shear-Wap [7]. Voxels represent the properties of a point in space, such as color, density, opacity, etc. In the actual calculation process, the voxel can be simplified into a cube, as shown in Figure 2.

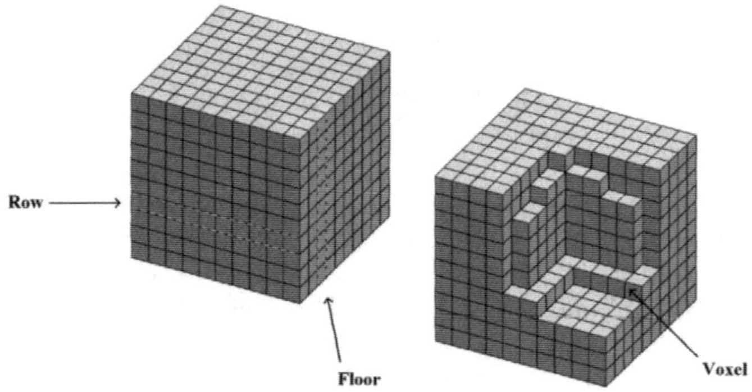

Figure 2. Volume data and voxel

3. Reconstruction process and result analysis

3.1. *Image denoising*

There are often a lot of noises in the two-dimensional images directly generated by ICT machine. Therefore, image denoising must be performed before image processing. The main purpose of image denoising is to improve the accuracy of the reconstructed model. The methods commonly used for processing are Disk filtering, Gaussian filtering, Average filtering, and Median filtering. During the denoising process, care must be taken not to destroy the edge information of the image, otherwise the edge information cannot be effectively extracted. The processing results of four different methods are shown in Figure 3. In the figure, a indicates the Disk filtering, b indicates the Gaussian filtering, and c indicates the Average filtering and d indicates the Median filtering.

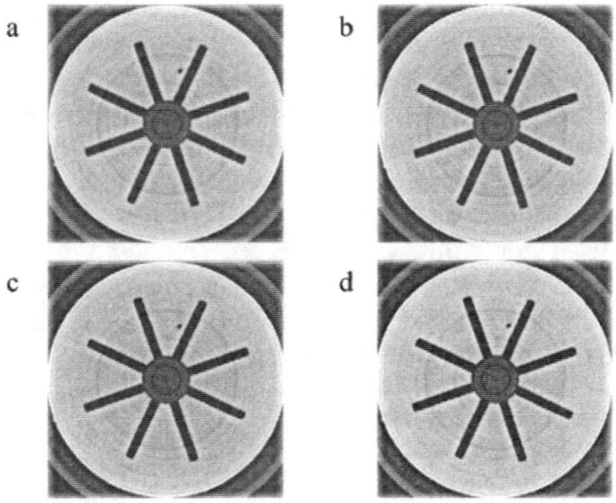

Figure 3. Denoising effect of different methods

As can be seen from the figure, different filtering methods have significant differences when filtering images. Disk filtering caused a blurred visual effect after processing, and also filtered some valid information in the image; Gaussian filtering and Average filtering produce clearer images, but during the processing process, the edges of some details are damaged, which will affect subsequent analysis; compared with the above three methods, Median filtering not only can effectively filter noise, but also has a better retention of image edge information.

3.2. Edge detection

The main principle of edge detection is to determine the gray level change rate of pixel information in a certain area of the image through algorithms. The change law of the direction derivative near the edge has obvious characteristics compared to other areas, so that the edge information of the image is obtained. The edge detection operators mainly include Roberts operator, Sobel operator, Prewitt operator, Approxcanny operator, Log operator and Canny operator. The test results of different operators are shown in Figure 4, and the test results are analyzed and compared in the following.

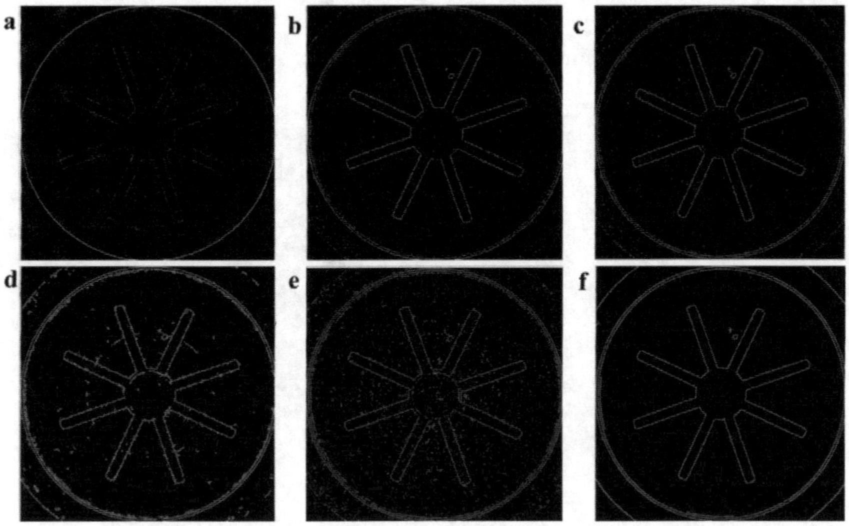

Figure 4. Test results obtained by different operators

IOP Conf. Series: Materials Science and Engineering **816** (2020) 012009 doi:10.1088/1757-899X/816/1/012009

In the figure, a indicates Roberts operator, b indicates Sobel operator, c indicates Prewitt operator, d indicates Approxcanny operator, e indicates Log operator, and f indicates Canny operator. From the test results in the figure, the results obtained by different operators are quite different. Roberts operator can get relatively thin edges, but the information is seriously lacking, resulting in discontinued edges; Sobel operator and Prewitt operator get continuous results, but it is easy to produce false edges; Approxcanny operator and Log operator have better continuity, but they are more sensitive to noise and cannot effectively identify noise; compared with other operators, the Canny operator obtains a clearer, complete and smooth edge, and can effectively recognize the noise contained in the image.

3.3. Gray-scale image segmentation

Image segmentation is one of the basic problems of image processing and computer vision. Segmentation is usually used for further analysis and recognition of images. The purpose of segmentation is to separate the target area and background area of the image, so as to further reduce the interference of background factors and improve the quality of image reconstruction. During the research, it was found that the selection of the threshold has a great impact on the results obtained. Figure 5 shows the effect of choosing inappropriate thresholds on the segmentation results. In the figure, a and b respectively show two cases where the threshold is too large and too small. When the threshold is selected too large, it is easy to overwhelm the contained defects and the SRM case is likely to affect the results; however, when the threshold is selected too small, it is easy to confuse some areas inside the grain with the background. Therefore, an appropriate threshold value needs to be selected to obtain an ideal segmentation result. Figure 6 shows that the histogram threshold method is used to determine the threshold, and the obtained threshold is used for image segmentation. A clear segmentation result can be observed from the figure.

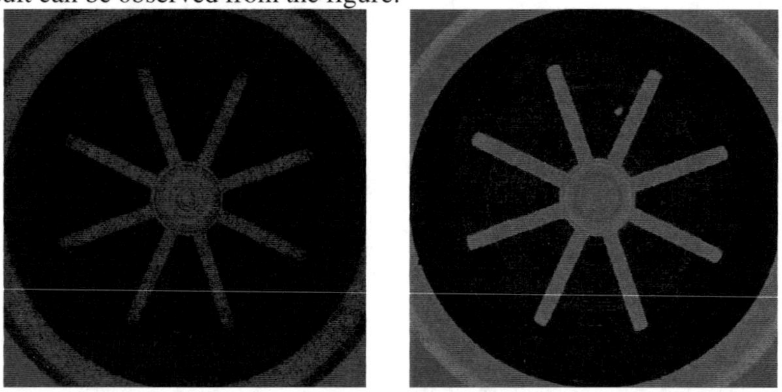

a **b**

Figure 5. Segmentation result of inappropriate threshold

Figure 6. Histogram threshold segmentation results

IOP Conf. Series: Materials Science and Engineering **816** (2020) 012009 doi:10.1088/1757-899X/816/1/012009

3.4. Reconstruction results display

The 3D model is obtained after rendering the 2D image processed as described above. But some noise is inevitably retained in the volume data, which causes the surface of the reconstructed model to be not smooth enough. Therefore, the surface of the model needs to be smoothed, and more realistic results can be obtained after the smoothing, as shown in Figure 7.

Figure 7. Reconstruction model

Sectioning the generated model can facilitate the feature extraction and classification of defects. The sectioning method is to find the intersection of the anatomical section and each slice in the 3D model according to the given section, divide the slice into two different regions through the intersection line, and redraw the segmented model to get the desired section model. Figure 8 shows the results of cutting along.

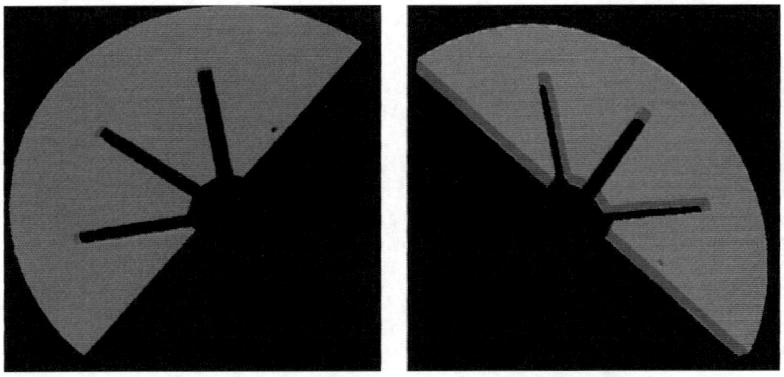

Figure 8. The results of cutting along

The Volume rendering image is essentially a linear superposition of the color values and transparency of the sampling points on the ray. If different color values are assigned to different structures, the opacity of the model can be adjusted when rendering, and then the internal structure and relative position can be observed. As shown in Figure 9.

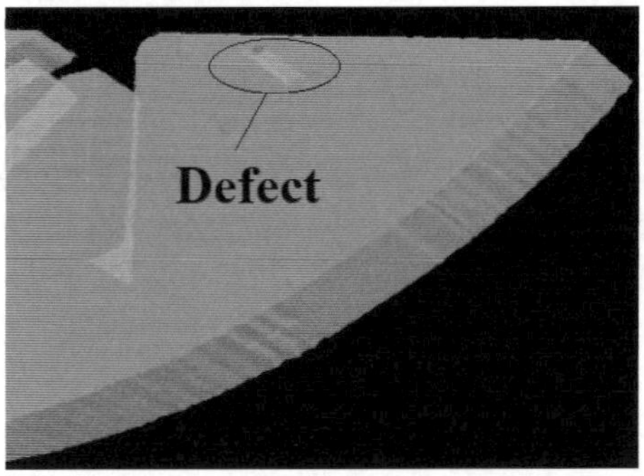

Figure 9. Reconstruction model

IOP Conf. Series: Materials Science and Engineering **816** (2020) 012009 doi:10.1088/1757-899X/816/1/012009

4. Summary

This paper studies the application of continuous tomographic ICT image 3D reconstruction technology in the non-destructive testing of defective SRM, and makes the following conclusions:

1. Briefly analyzed the basic ideas of surface rendering and volume rendering in 3D reconstruction , and explain the advantages and disadvantages of each method;
2. The effects of four different denoising algorithms are explained in comparison, and the effect of the median filtering algorithm is the most ideal;
3. Analyzed the effect of six different operators on image edge detection, and conclude that the effect of Canny operator is more clear than several other operators;
4. Analyzed the effect of threshold selection on gray-scale segmentation, and use the histogram threshold to obtain better segmentation results;
5. The results obtained in this paper can use the numerical simulation method to simulate the combustion and flow process of SRM, thereby getting some indicators to reflect the working performance of SRM.

5. References

[1] Jing Wang 2004 Complex-Product Oriented Industrial CT Image Reconstruction & Analysis *Northwestern Polytechnical University.*

[2] Aidong Zhang 2005 Three Dimensional Image Reconstruction of Industrial Computed Tomography based on continuous tomograms *China Academy of Engineering Physics.*

[3] Jingyang Han 2019 Research on 3D Visualization Technology of Dental CT Images *Henan University of Science and Technology.*

[4] Junhong Zhao 2003 Research on 3D Reconstruction and 3D Reveal with serial slices of Industrial Computerized Tomography *Chongqing University.*

[5] Fei Wang 2005 Research on the visualization of NDT of solid rocket engine based on ICT *Northwestern Polytechnical University.*

[6] Jinbo Zhang 2008 The Research of Reverse 3D CAD Modeling Based on Industrial CT Sectional Images *Chongqing University.*

[7] Yu Ma 2007 Reconstruction and Visualization of Edge Surfaces within 3D Images *Shanghai Jiao Tong University.*

[8] Qinggui Chen 2014 CT image reconstruction of solid rocket motor based on narrow fan beam scan. Journal of Solid Rocket Technology **37** 6:877-880.

[9] Guang Wang, Shusheng Zhang, Zhengxi Guan 1999 Research on the Inside and Outside Boundaries Recognition on ICT Image of Assembly Parts *Mechanical Science and Technology for Aerospace Engineering* 4:108-110.

[10] Yao Cheng, Guangmin Hu, Cai Lu 2012 Algorithm of Eliminating Illegal Border of Texture Based on the Volume Smooth *Journal of Network New Media* **1** 4:17-22.

[11] Rongzhen Liu 2008 Industrial CT Measurement Technology to Solid Rocket Engine *Tactical Missile Technology* 5:92-96.

[12] Hongyi Lu, Guiqing Chen, Hongmei Zhou 2016 Streak artifacts correction of solid rocket motor's CT image *Journal of Aerospace Power* **31** 10:2515-2521.

[13] Junhong Zhao, Zhong Qu 2003 3D Reconstruction of 2D ICT Image *Microcomputer Development* 7:20-21+50.

[14] Min Zhu, Hongyi Lu 2014 Precise Measurement Technologies of Volume Space Defects Characters in Solid Rocket Motor. *Modern Defence Technology*, **42**, 1:117-122.

[15] Wencai Zhao 2010 3-D Reconstruction of High Energy X-Ray Film Images of the SRM. Nondestructive Testing **32** 6:460-462+474.

IOP Conf. Series: Materials Science and Engineering **816** (2020) 012010 doi:10.1088/1757-899X/816/1/012010

Aerodynamic Coefficients Prediction for 122 mm Rocket by Using Computational Fluid Dynamics

N L Tun[1], M Thuri[1], Minnzinoo[1] and Z Li[2]

[1] Department of Mechanical Engineering, Defence Services Technological Academy, Wetwun, Myanmar
[2] School of Energy and Power Engineering, Beihang University, Beijing, China

zeyulee@buaa.edu.cn

Abstract. In this study, the Computational Fluid Dynamics code in SolidWorks was used to predict the aerodynamic coefficients of 122 mm rocket in order to get more accurate aerodynamic coefficients and derivatives for flight simulation. This paper presented the governing equations of fluid dynamics applied in SolidWorks and developed the computational model of the rocket. The results were compared against calculated data from an empirical method. The conducted analyses showed that the comparison results are nearly the same. However, the empirical method provided a faster way to obtain results.

1. Introduction

This study focuses on the aerodynamics calculation and analysis of small rockets. The parameters of airflow characteristics, such as the drag force, pressure, and airflow velocity, have significant influence on the exterior ballistics characteristics of the rocket, and thus affect the effect and accuracy of the rocket. The essential thrust characteristics of the rocket engine are directly related to the drag coefficient and the drag force, which are the foremost aerodynamic parameters in the study of exterior ballistics. Computational fluid dynamics (CFD) is a currently pervasive application in aviation for evaluating aerodynamic performance during the conceptual stage and preliminary design stage in order to reduce design cycle time and minimize the experimental validation associated costs [1]. The purposes of this paper are to compare the aerodynamic coefficients of the static stability coefficient and the axial force coefficient of 122 mm rocket obtained by SolidWorks and an empirical method and to get more accurate aerodynamic coefficients and derivatives for flight simulation.

2. Empirical Method

2.1. Axial Force Coefficient Prediction

Drag has two fundamental sources of pressure distribution and surface friction around the rocket. Pressure distribution is further divided into base drag, parasitic pressure drag attributed to prominences such as emitter lug and body pressure drag including supersonic shock wave. The total axial force applied to the rocket can be described by:

$$F_x = 1/2 C_A \rho V^2 A_{ref} \tag{1}$$

Where C_A is the axial force coefficient, V is the absolute velocity, and A_{ref} is the reference area.

2.1.1. Skin Friction Drag. Skin friction is from the viscous air-flow around the rocket.

The skin friction drag coefficient for turbulent flow is calculated by [2, 3]:

$$C_f = 1.48\times10^{-2} \quad (R<10^4) \tag{2}$$

$$C_f = (1.5\ln R - 5.6)^{-2} \quad (10^4 < R < R_{crit}) \tag{3}$$

$$C_f = 0.032\left(R_s/L_{ref}\right)^{1/5} \quad (R > R_{crit}) \tag{4}$$

$$R_{crit} = 51\left(R_s/L_{ref}\right)^{-1.039} \tag{5}$$

Where R_s represents the surface roughness in μm , and L_{ref} represents the reference length in meter. The turbulent skin friction drag coefficient is corrected for compressibility with: [2, 3]

$$C_{fc} = \begin{cases} C_f\left(1-0.09M^2\right)(Subsonic) \\ C_f\left(1+0.15M^2\right)^{-0.58} \quad (Supersonic) \end{cases} \tag{6}$$

Where M represents the Mach number.
The skin friction drag coefficient based on the reference area can be found by [2, 3]

$$C_{Df} = C_f \cdot A_w / A_{ref} \tag{7}$$

Where A_w represents the wetted area of the rocket.
The wetted area of fins is twice its planform area, the skin friction drag coefficient of the tail is: [2, 3]

$$C_{Dft} = 2nC_{fc} \cdot A_w / A_{ref} \tag{8}$$

2.1.2. Pressure Drag of Fins. The pressure drag of fins depends on the fin contour shape, which has three typical shapes as a rectangular contour, a tapering trailing edge and circular leading edge, and a circular leading and trailing edges. The drag at the circular leading edge can be regarded as a cylinder without base drag in cross flow. Barrowman deduced an empirical formula as [2]

$$(C_D)_{LE\perp} = \begin{cases} -1+\left(1-M^2\right)^{-0.417} & (M<0.9) \\ 1\text{-}1.785(M-0.9) & (0.9<M<1) \\ 1.214-0.502/M^2+0.1095/M^4 & (M>1) \end{cases} \tag{9}$$

The subscript $\perp$ symbolizes that the leading edge is perpendicular to the flow.
The pressure drag at the leading edge of a slanted fin is derived from the principle of cross flow as [2]

$$(C_D)_{LE} = (C_D)_{LE\perp} \cdot \cos^2\Gamma_L \tag{10}$$

Where Γ_L represents the leading edge angle. Note that both coefficients in the equation are relative to the face area of the cylinder, so their reference area ratio is still $\cos\Gamma_L$.
The fin base drag coefficient of the square profile fin is equal to the body base drag coefficient: [2]

$$(C_D)_{TE} = (C_D)_{base} \tag{11}$$

The total fin pressure drag represents the sum of the trailing edge drag and the leading edge drag: [2]

$$(C_D)_{pressure} = (C_D)_{LE} + (C_D)_{TE} \tag{12}$$

2.1.3. Body Pressure Drag. The pressure drag is generated by the air around the rocket. This paper describes the estimation of the pressure drag of the nose cone, ignoring other body pressure resistance. The pressure drag of the streamlined nose cone is notably less than the skin friction drag at subsonic speeds. Figure 1 presents the measured pressure drag coefficients for different nose cone shapes.

The pressure drag of nose cones at low subsonic speed can be calculated by Equation (13): [2]

$$\left(C_{D,M=0}\right)_p = 0.8\sin^2\Phi \tag{13}$$

In this paper, a simple semi-empirical method is used to estimate the transonic pressure drag of the nose cone, rather than the general segmentation calculation.

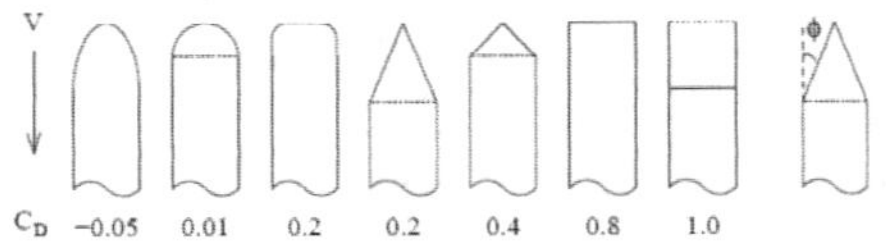

Figure 1. Pressure drag of various nose cone shapes.

This paper adopts a semi-empirical method under the circumstance of Mach number over 0.8. The pressure drag is interpolated between the discreet value of Equation (13) and the transonic method at high subsonic speeds. Assume that the pressure drag does not decrease in the subsonic zone, and the derivative is zero at $M=0$. An interpolation function similar to the Prandtl factor is proposed as: [2]

$$\left(C_D\right)_{pressure} = aM^b + \left(C_{D,M=0}\right)_p \tag{14}$$

Where a and b are calculated for the fitting analysis of the drag coefficient as well as its derivative at the lower limit of the transonic method.
Equation (15) is to calculate the pressure drag of the conical nose cone at supersonic speeds (M > 1.3). The drag between Mach 1 and 1.3 is interpolated with polynomials using the boundary conditions:

$$\left(C_D\right)_{pressure} = 2.1\left(1+4f_N^2\right)^{-1} + 0.5\left[\left(1+4f_N^2\right)\left(M^2-1\right)\right]^{-1/2} \tag{15}$$

$$\tan\varepsilon = 1/\left(2f_N\right) \tag{16}$$

$$\partial\left(C_D\right)_{pressure}/\partial M = 4\left(\gamma+1\right)^{-1}\left(1-0.5C_{D,M=1}\right) \tag{17}$$

Where the drag coefficient is approximately ε at $M=1$ and γ=1.4 is the air specific heat ratio.

$$C_{D,M=1} = \sin\varepsilon \tag{18}$$

At supersonic and transonic speeds, the pressure drag of ogival nose cones with identical length with shape factor correction and diameter is equal to that of conical nose cones as:

$$\left(C_D\right)_{pressure} = \left[0.72\left(\kappa-0.5\right)^2 + 0.82\right]\left(C_D\right)_{cone} \tag{19}$$

The shape factor is 1 at $\kappa=0.1$ and 0.82 at $\kappa=0.5$. For ogival nose cone, the parameter κ is one.

2.1.4. Base Drag. Base drag is generated by anywhere where the body diameter rapidly decreases or the low pressure area at the bottom of the rocket. An empirical formula can be used to estimate the magnitude of the base resistance as: [2]

$$\left(C_D\right)_{base} = \begin{cases} 0.12+0.13M^2 & (M<1) \\ 0.25/M & (M>1) \end{cases} \tag{20}$$

A reasonable approximation is obtained by reducing the propulsion motor area from the reference region [2]. Therefore, the base drag will not be generated with the same base size. In addition, with a large base and a single small motor at the center, the basic drag is roughly the same as when sliding.

2.1.5. Total Drag. The pressure drag is generated by the air around the rocket. This paper describes the estimation of the pressure drag of the nose cone, ignoring other body pressure resistance.

$$C_D = C_{Df} + C_{Dft} + \left(C_D\right)_{pressure} + \left(C_D\right)_{base} \tag{21}$$

2.2. Static Stability Coefficient Prediction

A rocket is regarded to be: [4]

- Statically unstable if the disturbance will keep the rocket away from equilibrium.
- Statically stable if small perturbations tend to bring the rocket back to equilibrium.

Static stability is demonstrated in figure 2. The center of gravity (CG) is the point where the rocket rotates about when disturbed. The center of pressure (CP) is the point where synthetic aerodynamic forces are assumed to act. Note that CG and CP may not be on the longitudinal centerline of the rocket. In this case, the effect of axial force on stability should be considered. In short, a rocket can be considered statically stable when CP is behind CG, both measured from the rocket tip. [4].

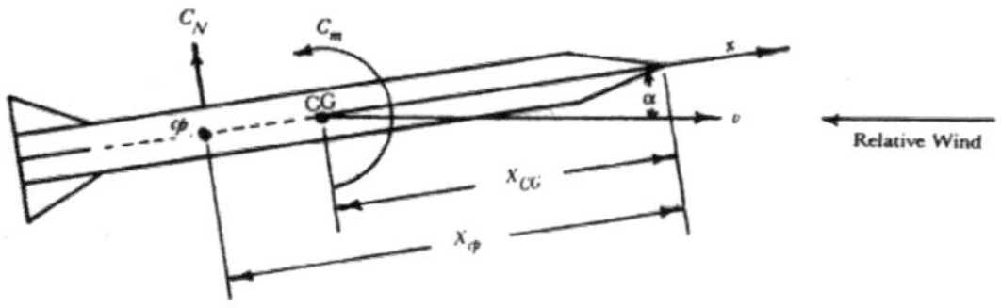

Figure 2. C_N, C_m and α relationships.

2.2.1. Normal Force. Barrowman proposed formulas to estimate the normal force and the position of CP on a small rocket. The Barrowman equations are derived on the basis of the assumptions: [5]

- The compressibility effect is negligible, i.e. $M_a < 0.4$.
- The rocket has a low Angle of attack ($\alpha < 10°$).
- The lift on the rocket body tube is negligible.
- The viscous forces can be neglected.
- The nose of the rocket is poised at a point.
- The rocket is relatively narrow in diameter.
- Steady air flow over the rocket without rapid changes.

The normal force on the rocket can be calculated by:

$$F_N = 0.5 C_N \rho V^2 A_{ref} \tag{22}$$

Where, C_N is the normal force coefficient.

At very small attack angles, C_N can be approximated as linear as α . Hence, it can be expressed as:

$$C_N \approx C_{N\alpha}\alpha \tag{23}$$

Where α represents the attack angle in radian and $C_{N\alpha}$ represents the slope of the normal force coefficient, also known as derivative of stability.

The total derivative of stability can be derived from the sum of the derivatives of stability for each part.

$$C_{N\alpha(R)} = \sum_{P\in R} C_{N\alpha(P)} \tag{24}$$

Where R represents the whole rocket and P represents a rocket part.

The CP position of the rocket is calculated according to CP of every independent rocket component. Similar to the center of mass (CM), CP locates on the rocket's rolling axis, the CP location is defined as the distance (X_{CP}) from CP at the tip of the nose cone. X_{CP} can be calculated using Equation (25).

$$X_{CP(R)} = \sum_{P\in R} C_{N\alpha(P)} X_{CP(P)} / C_{N\alpha(R)} \tag{25}$$

Where $X_{CP(P)}$ is the distance from CP of component P to the tip of the nose cone of the rocket.

Equations of $C_{N\alpha}$ and X_{CP} for conical variations in body diameter, fins and nose cones are as follows. As long as the nose cone is parabolic, conical or ogive, the stability derivative of the nose cone is two.

$$C_{N\alpha(n)} = 2 \tag{26}$$

The position of the center of pressure for the nose cone depends on the shape:

$$X_{CP(n)} = \begin{cases} 0.667l_n & (Conical) \\ 0.466l_n & (Ogive) \\ 0.5l_n & (Parabolic) \end{cases} \tag{27}$$

Where l_n is the length of the nose cone and $X_{CP(n)}$ is the distance from CP to the tip of the nose cone. The derivative of stability of a trapezoidal fin can be calculated by the equation as:

$$C_{N\alpha(f)} = K_{fb} \cdot 4nl_s/d_n \left\{1 + \left[1 + (2l_m/l_r + l_t)^2\right]^{1/2}\right\}^{-1} \tag{28}$$

Where l_s represents the fin span, l_f represents fin root-chord, l_m represents fin mid-chord, l_t represents fin tip-chord and n represents fin number. K_{fb} represents the coefficient of increased normal force due to the interference effect between the fin and the body.

$$K_{fb} = 1 + d_f\left(2l_s + d_f\right)^{-1} \tag{29}$$

Where d_f represents the body tube diameter at the position of fin.
The position of CP for a trapezoidal fin can be calculated by:

$$X_{CP(f)} = X_f + l_m(l_r + 2l_t)\left[3(l_r + l_t)\right]^{-1} + 1/6\left[l_r + l_t - l_r l_t (l_r + l_t)^{-1}\right] \tag{30}$$

Where X_f represents the length between the tip of the nose cone and the intersection of fin leading edge and body tube.
The method described above for calculating C_N is only valid for incompressible flow. It is shown how to extend the range of the aerodynamic coefficient into a compressible flow state.
For fin component, the Prandtl-Glauert compressibility correction can be applied as: [5]

$$C'_{N\alpha(f)} = \begin{cases} C_{N\alpha(f)}\left(1 - M^2\right)^{-1/2} & (M < 0.8) \\ C_{N\alpha(f)}\left(1 - 0.8^2\right)^{-1/2} & (0.8 \le M \le 1.1) \\ C_{N\alpha(f)}\left(M^2 - 1\right)^{-1/2} & (M > 1.1) \end{cases} \tag{31}$$

For the slender body, Ericsson compressibility correction is applied: [6]

$$\left(C_{N\alpha}\right)_b = 2K_m \tag{32}$$

The term K_m is a Mach correction factor that is empirically derived from the body alone data:

$$K_m = \begin{cases} 1 - 0.23\beta^2 & (M < 1) \\ -0.130006\beta^2 + 0.627\beta + 1 & (M \ge 1) \end{cases} \tag{33}$$

Where,

$$\beta = \left(M^2 - 1\right)^{1/2} \tag{34}$$

2.2.2. Pitching Moment Coefficient. The pitching moment coefficient can be calculated as:

$$C_m = M_z\left(0.5\rho V^2 A_{ref} L_{ref}\right)^{-1} \tag{35}$$

At very small angles of attack, C_m can be approximated as linear as α. Hence, it can be expressed as:

$$C_m \approx C_{m\alpha}\alpha \tag{36}$$

Where $C_{m\alpha}$ is the slope of the pitching moment coefficient.

The static stability degree is a function of the $\frac{C_{N\alpha}}{C_{m\alpha}}$. The algebraic symbol of $C_{m\alpha}$ indicates whether the rocket is statically unstable or stable, and the magnitude of $C_{m\alpha}$ measures the stability degree.
The margin of stability can be defined as:

$$C_{m\alpha} = C_{N\alpha} \cdot (X_{CG} - X_{CP})/d \tag{37}$$

Where X_{CG} is the center of gravity from the nose cone tip.

3. Solidworks Software

SolidWorks Flow Simulation is a new form of CFD analysis software totally embedded in the environment of mechanical design, suitable for all kinds of general engineering applications. [7]
SolidWorks Flow Simulation is based on a Cartesian grid, which is one of the essential elements of the CAD/CFD Bridge for CAD-embedded CFD. As a result of the Cartesian grids application, cells are entirely located in fluid cells, solid cells and partial cells. In the simplest case, the partial cell consists of 2 control volumes (CV): a solid CV and a fluid CV, as shown in figure 3.

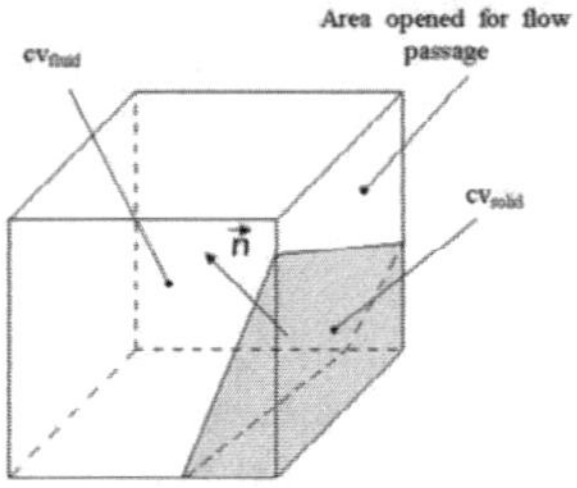

Figure 3. A partial cell in the simplest case and with 2 control volumes (CV) inside.

3.1. Background Theory of SolidWorks

SolidWorks Flow Simulation solves the Navier-Stokes equations. [7]
Continuity equation:

$$\frac{\partial \rho}{\partial t} + \frac{\partial(\rho u_i)}{\partial x_i} = 0 \tag{38}$$

Where u represents the vector of velocity and ρ represents the density.
Momentum equation:

$$\frac{\partial(\rho u_i)}{\partial t} + \frac{\partial(\rho u_i u_j)}{\partial x_i} + \frac{\partial P}{\partial x_i} = \frac{\partial(\tau_{ij} + \tau_{ij}^R)}{\partial x_i} + S_i \tag{39}$$

Where P is the pressure, S is the momentum source, τ is the stress tensor.
Energy equation:

$$\frac{\partial(\rho H)}{\partial t} + \frac{\partial(\rho u_i H)}{\partial x_i} + \frac{\partial P}{\partial x_i} = \frac{\partial\left[u_j(\tau_{ij} + \tau_{ij}^R) + q_i\right]}{\partial x_i} + \frac{\partial P}{\partial t} - \tau_{ij}^R \frac{\partial u_i}{\partial x_j} + \rho\varepsilon + S_i u_i + Q_H \tag{40}$$

Where H is the total enthalpy, related to the static enthalpy h by:

$$H = h + u^2/2 \tag{41}$$

The following energy equation is adopted for high-speed compressible flows and shock wave flows as:

$$\frac{\partial(\rho E)}{\partial t} + \frac{\partial\left[\rho u_i (E + P\rho^{-1})\right]}{\partial x_i} = \frac{\partial\left[u_j(\tau_{ij} + \tau_{ij}^R) + q_i\right]}{\partial x_i} + \frac{\partial P}{\partial t} - \tau_{ij}^R \frac{\partial u_i}{\partial x_j} + \rho\varepsilon + S_i u_i + Q_H \tag{42}$$

$$E = e + u^2/2 \tag{43}$$

IOP Conf. Series: Materials Science and Engineering **816** (2020) 012010 doi:10.1088/1757-899X/816/1/012010

3.2. Modeling and Mesh Generation of 122 mm Rocket

Note that the optimal size of the control volume is 5 to 10 times the length of model at the side and back and at least the length of model in the front [8]. In the paper, the control volume is 5 times the model length at the side and back, and one length in the front of the rocket.

Geometric characteristics of the 122 mm rocket is shown in figure 4.

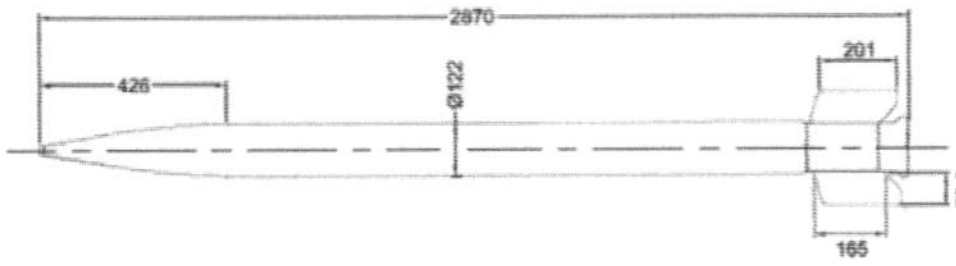

Figure 4. Geometric characteristics of 122 mm rocket.

The SolidWorks models of the 122 mm rocket are shown in figure 5.

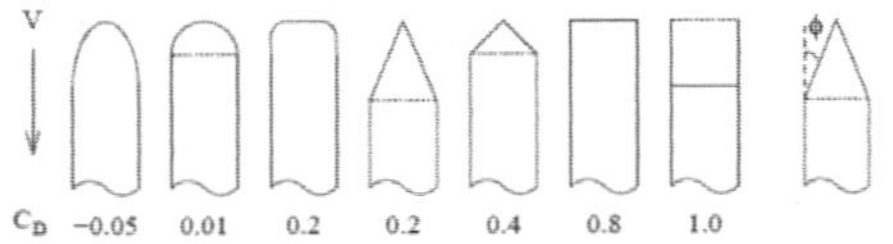

Figure 5. SolidWorks model for 122 mm rocket with computational domain.

In the flow analysis, numbers of divisions are chosen 85 along X-axis and 75 along Y, Z-axis for initial mesh generation in Global Mesh setting. Local Mesh is added for the entire rocket to increase grid density. Grid generation on the 122 mm rocket is shown in figure 6.

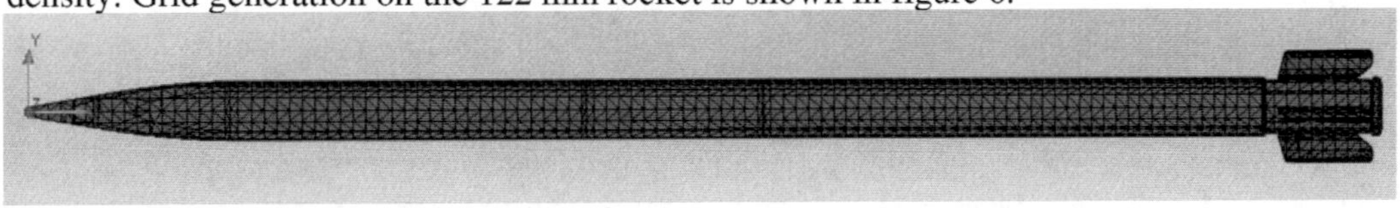

Figure 6. Geometric characteristics of 122 mm rocket.

4. Result and Discussions

Simulation data were obtained according to various Mach numbers ranging from *M= 0.2* to *M=2.2* with a step of *0.2* at *0°* and *2°* angles of attack. The computer to run SolidWorks Flow Simulation adopted a Quad-core Intel® Core™ i7-6500U processor running at 2.5 GHz with 12GB of RAM. The computer ran the 64-bit version of SolidWorks 2017. The average running time of one data point was approximately *1.5* hours.

The airflow velocity and pressure distribution contours on the cross-section plane, when the Mach number is one at *0°* angles of attack are shown in figure 7 and figure 8 respectively.

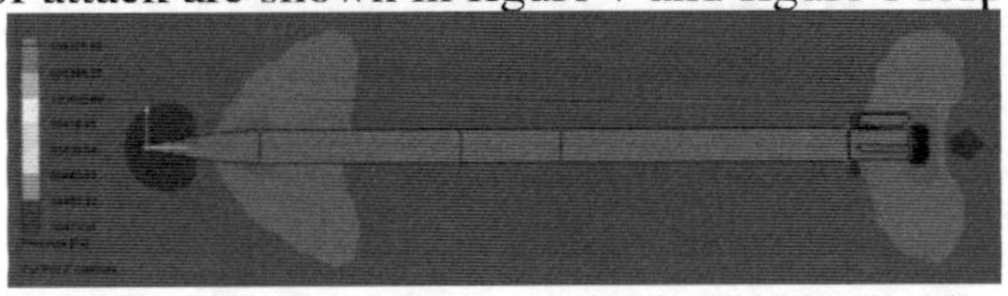

Figure 7. Pressure contour plot in XY body plane.

High pressure regions occur on the rocket nose and fin leading edges due to air wave. Low pressure region occurs at the rocket base due to airflow separation from the rearward-facing step. These can be seen in figure 7. Due to low pressure condition, the air flow velocity is decreased behind the rocket base of the low pressure region. It can be seen in figure 8.

Figure 8. Pressure contour plot in XY body plane.

F_x and F_y obtained from the simulations represents the normal and axial force respectively. F_x and F_y are equivalent to the drag force (F_D) and lift (F_L) respectively at zero attack angle. For other attack angles, F_x and F_y are not equal to F_D and F_L as shown in figure 9. F_L and F_D will be obtained through further calculations, as shown in Equation (44), Equation (45) and Equation (46). [9]

$$F_D = F_x \cos\alpha + F_y \sin\alpha \tag{44}$$

$$F_L = F_y \cos\alpha - F_x \sin\alpha \tag{45}$$

$$F_N = F_L \cos\alpha + F_D \sin\alpha \tag{46}$$

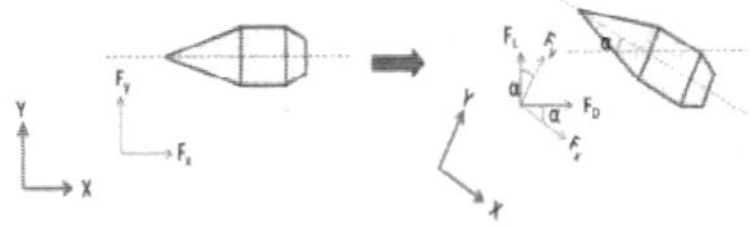

Figure 9. Forces acting on the projectile.

The results for C_N and C_M at various angles of attack are calculated using Equation (22) and (35), while $C_{N\alpha}$ and $C_{M\alpha}$ at various Mach numbers are calculated using Equation (47) and (48). In this paper, *0°* and *2°* are used for the computation for $C_{N\alpha}$ and $C_{M\alpha}$. [9]

$$C_{N\alpha} = \left(C_{N\alpha_2} - C_{N\alpha_1}\right)\left(\alpha_2 - \alpha_1\right)^{-1} \tag{47}$$

$$C_{M\alpha} = \left(C_{M\alpha_2} - C_{M\alpha_1}\right)\left(\alpha_2 - \alpha_1\right)^{-1} \tag{48}$$

The comparison between the empirical and CFD computed axial force coefficient shows a difference of about *31* percent in the subsonic range, good agreement in the transonic region, and a difference increases in start point of the supersonic region, but a difference is gradually decreasing for increasing Mach number, as was shown in table 1 and figure 10.

Table 1. Axial force coefficient calculated with CFD and empirical method.

Mach Number	0.2	0.4	0.6	0.8	1.0	1.1	1.2	1.4	1.6	1.8	2.0	2.2
CFD	0.46	0.48	0.50	0.54	0.60	0.61	0.61	0.57	0.54	0.51	0.47	0.42
Empirical	0.28	0.32	0.38	0.47	0.59	0.59	0.59	0.48	0.45	0.42	0.40	0.38

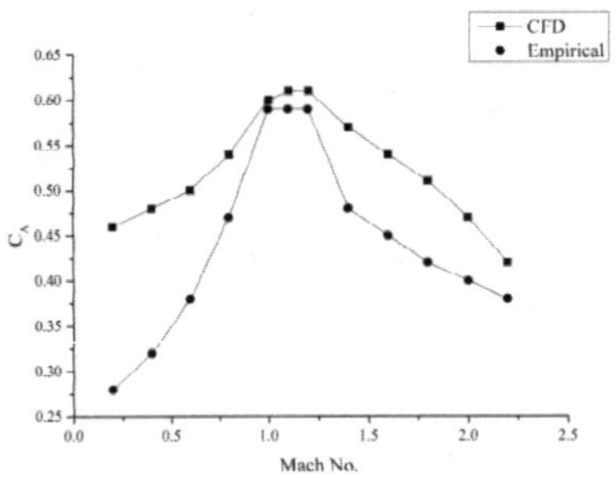

Figure 10. Comparison of axial force coefficient.

The comparison of the normal force coefficient and normal force coefficient derivative calculated with CFD and Empirical method shows the good agreement in subsonic and transonic region, but a different of about *13%* and *15%* at *M=0.8* and *M=1.1*, respectively. The results are shown in table 2 and table 3. Moreover, a difference in the supersonic region is a little more than in other regions. These are illustrated in figure 11 and figure 12.

Table 2. Normal force coefficient calculated with CFD and empirical method.

Mach Number	0.2	0.4	0.6	0.8	1.0	1.1	1.2	1.4	1.6	1.8	2.0	2.2
CFD	0.24	0.24	0.24	0.25	0.27	0.26	0.26	0.25	0.25	0.22	0.22	0.21
Empirical	0.21	0.22	0.24	0.29	0.28	0.30	0.28	0.23	0.20	0.19	0.18	0.17

Table 3. Normal force coefficient derivative calculated by CFD and empirical method.

Mach Number	0.2	0.4	0.6	0.8	1.0	1.1	1.2	1.4	1.6	1.8	2.0	2.2
CFD	7.09	7.02	7.12	7.35	7.80	7.52	7.44	7.33	7.14	6.53	6.53	6.27
Empirical	6.24	6.45	6.94	8.37	8.20	8.66	8.24	6.65	5.98	5.59	5.32	5.11

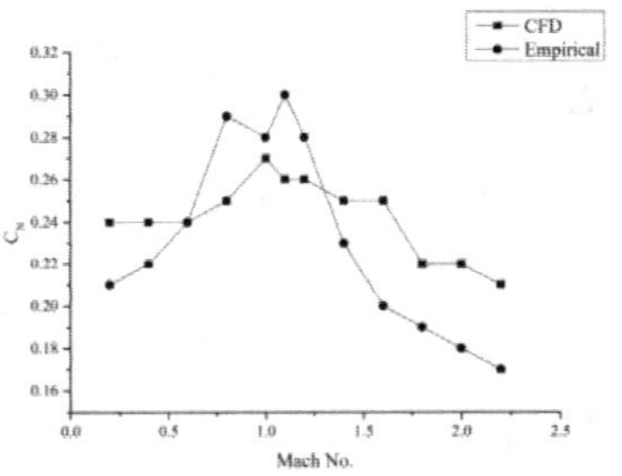

Figure 11. Comparison of normal force coefficient.

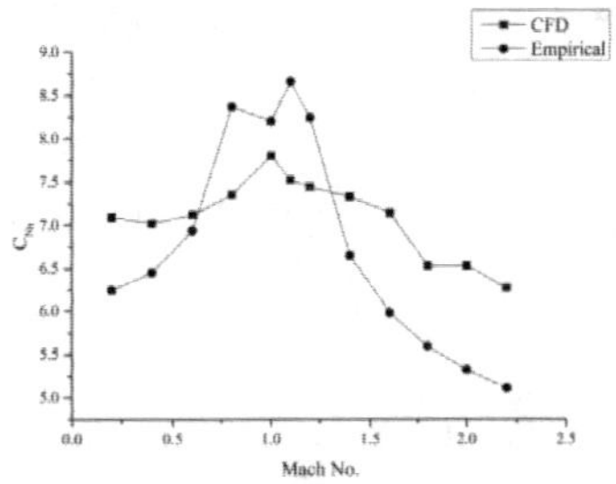

Figure 12. Comparison of normal force coefficient derivative.

The comparison of the pitching moment coefficient and pitching moment coefficient derivative calculated with CFD and Empirical method shows the good agreement in subsonic, but with a little difference in transonic range, and the difference is increasing in the supersonic region. These are illustrated in table 4, table 5, figure 13 and figure 14.

Table 4. Pitching moment coefficient calculated with CFD and empirical method.

Mach Number	0.2	0.4	0.6	0.8	1.0	1.1	1.2	1.4	1.6	1.8	2.0	2.2
CFD	-0.99	-1.18	-1.21	-1.32	-1.49	-1.23	-1.15	-1.01	-0.93	-0.70	-0.67	-0.52
Empirical	-1.03	-1.07	-1.15	-1.39	-1.36	-1.44	-1.37	-1.10	-0.99	-0.92	-0.88	-0.84

Table 5. Pitching moment coefficient derivative calculated by CFD and empirical method.

Mach Number	0.2	0.4	0.6	0.8	1.0	1.1	1.2	1.4	1.6	1.8	2.0	2.2
CFD	-28.56	-33.97	-34.86	-37.94	-42.72	-35.34	-33.06	-28.93	-26.83	-20.09	-19.27	-15.04
Empirical	-29.70	-30.69	-33.06	-39.83	-39.04	-41.24	-39.23	-31.66	-28.46	-26.60	-25.31	-24.30

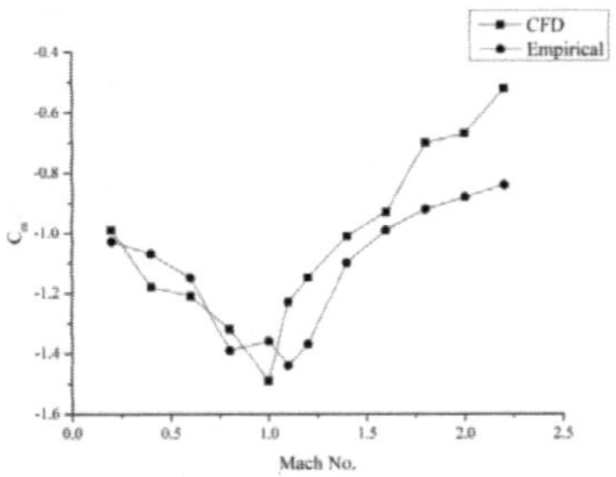

Figure 13. Comparison of pitching moment coefficient.

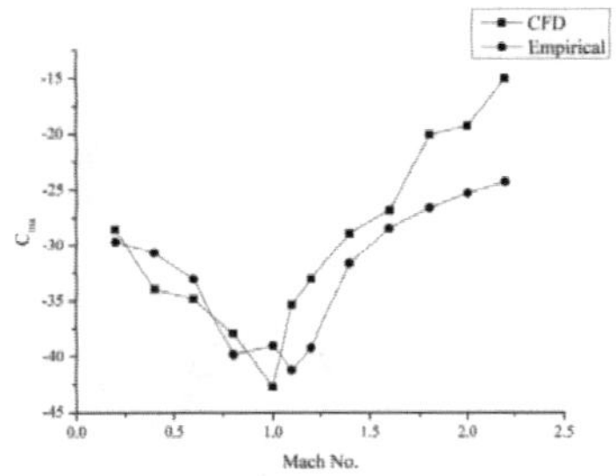

Figure 14. Comparison of pitching moment coefficient derivative.

5. Conclusions

The conducted analyses have shown that the comparison results are nearly the same. For the axial force coefficient, the differences are apparent in the subsonic region. However, for static stability coefficients and derivatives, the differences are noticeable in the supersonic region. For all coefficients, the comparison results show a good agreement in the transonic region.

Generally, although more accurate results can be achieved by CFD method, the empirical method has provided a very fast way to obtain results. Based on the numerical solution of the Navier-Stokes equations, the CFD code took about *1.5* hours for one data point.

6. References

[1] Tan W C and Sor W L 2010 *Aerodynamic validation of emerging projectile and missile configurations* (Monterey: Naval Postgraduate School Press)

[2] Niskanen S 2009 *Development of an Open Source model rocket simulation software* (Helsinki: Helsinki University of Technology Press)

[3] Barrowman J S 1967 *The practical calculation of the aerodynamic characteristics of slender finned vehicles* (Washington, DC: Catholic Univ. of America Press)

[4] Command U A M 1968 *Design of Aerodynamically Stabilized Free Rockets* (Alabama: U. S. Army Aviation and Missile Command)

[5] Box S, Bishop C M and Hunt H 2009 *Estimating the dynamic and aerodynamic parameters of passively controlled high power rockets for flight simulation*. (Cambridge: Cambridge Rocketry)

[6] Heiser M M 2017 *Dynamic Stability and Control: Methods and Developments* (Alabama: Auburn University Press)

[7] Sobachkin A and Dumnov G 2013 *Numerical basis of CAD-embedded CFD* (Salzburg: NAFEMS World Congress) p 9

[8] Wee H C 2011 *Aerodynamic analysis of a canard missile configuration using ANSYS-CFX* (Monterey: Naval Postgraduate School Press)

[9] Wong C H 2011 *Aerodynamic Analysis of the M33 Projectile Using the CFX Code*. (Monterey: Naval Postgraduate School Press)

Acknowledgments

The authors thank Dr Thein Oo, Dr Myint Win, and A.P. Haiying Lin for expert guidance and technical assistance. This study is supported by D.S.T.A., and receives assistance from Beihang University.

Research Progress of Anti-jamming Technology of Unmanned Aerial Vehicle(UAV) Data Link

Guo Shichao[1], Guo Dandan[2], Zhang Qiongyu[3], Wu Nankai[4], Deng Jiaxin[5]

[1,3,4,5]Beijing Institute of Spacecraft Environment Engineering, Beijing, China
[2]Beijing Institute of Spacecraft System Engineering, Beijing, China

Email: 357464199@qq.com

Abstract. An high-speed stealth UAV will be developed by an aviation research institute. Data link is an important part of UAV system. In order to have a deeper understanding of data link system, this paper was made through studying many recent academia papers. The basic composition and working principle of UAV data link system was introduced in this paper, the current UAV data link interference and anti-interference technology was expounded, the key problems faced by the development of UAV data link anti-interference technology were summarized, and the future development trend of UAV data link anti-interference technology was pointed out.

1. Overview of UAV data link system

These guidelines, written in the style of a submission to *J. Phys.: Conf. Ser.*, show the best layout for your paper using Microsoft Word. If you don't wish to use the Word template provided, please use the following page setup measurements.

1.1. basic composition

UAV data link system is usually composed of ground system and airborne system. Among them, the ground system commands, controls, tracks and positions the UAV in the air through wireless transmission [3], and its framework is shown in Figure 1. UAV data link system is divided into uplink and downlink. The uplink transmits the command of the ground system to the UAV; the downlink transmits the telemetry data and infrared or television images of the UAV to the ground system [4,5].

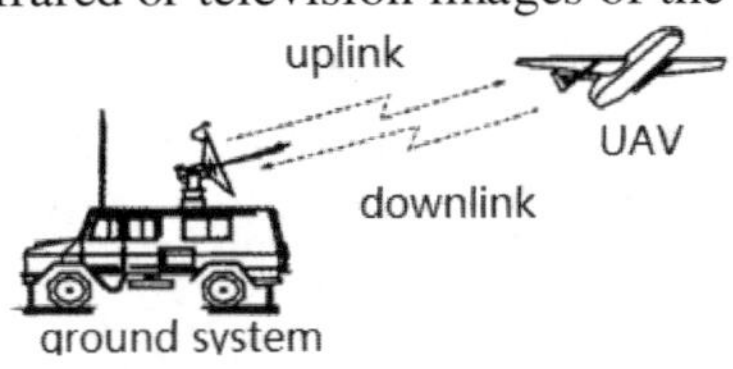

Figure 1. Basic composition of UAV data link system

1.2. Working principle

At the transmitting end, the ground system generates the remote control command, which is sent to the ground data terminal for coding, encryption and spread spectrum to form the remote baseband signal; then through the transmitter for phase modulation, it generates the remote intermediate frequency signal, which is converted up to form the remote radio frequency signal; then after power amplification, it is output to the remote transmission antenna, and then transmitted to the UAV through the ground air link. At the receiving end, the UAV receives the remote control RF signals through the antenna and the receiver respectively, and sends the remote control instructions to the UAV flight control system after amplification, despreading, demodulation and other processing [6].

2. Data link interference and anti-interference technology

Please follow these instructions as carefully as possible so all articles within a conference have the same style to the title page. This paragraph follows a section title so it should not be indented.

2.1. data link interference

Data link interference refers to the interference of electromagnetic wave emitted by radio interference equipment on data link receiving function of UAV. Typical UAV data link interference includes monophonic interference, modulation interference, broadband noise interference, sweep interference and other interference. Single tone interference and modulation interference belong to narrow-band interference; modulation interference is to modulate narrow-band noise to DS-SS working frequency point, including noise AM interference and noise FM interference; wide-band noise interference and sweep interference belong to wide-band interference, whose essence is to increase the background noise of the system; sweep interference is linear frequency modulation(LFM) signal, sweeping in the working frequency band of the system [7].

From the perspective of implementation, data link interference can be divided into suppressive interference and deceptive interference. Oppressive jamming is to shield the spectrum signal of enemy data link by transmitting some jamming signal, so that the received signal of enemy data link is blurred or completely covered. According to the width of jamming channel, the jamming can be divided into aiming jamming, semi aiming jamming and blocking jamming. Deceptive jamming is to make enemy drones execute our instructions by imitating enemy data link signals. There are two ways to implement deceptive jamming: one is to intercept and analyze the enemy data link signal, and then implement jamming; the other is to detect and acquire the enemy data link frequency band, copy its data link signal, and then send it to the enemy data link equipment to implement jamming [8].

2.2. Data link anti-jamming technology

At present, the anti-jamming technology of UAV data link mainly includes channel coding technology, spread spectrum technology, multiple input and multiple output technology, orthogonal frequency division multiplexing technology, adaptive antenna technology and perception radio technology [9,10].

2.2.1.channel coding technology. When the data link of UAV is disturbed, there will be error code in the transmission data, so the remote control command, telemetry parameter or image received by the receiver will be wrong. According to Shannon's channel coding theory, as long as the transmission rate of information is lower than the channel capacity, a channel coding method can be found to reduce the error probability arbitrarily. Therefore, the anti-interference ability of data link can be improved by processing data through channel coding [11]

2.2.2.spread spectrum technology. Spread spectrum technology is to *modulate* the transmitted signal to a wide frequency band, and recover the signal to the information bandwidth through correlation reception at the receiving end [12]. According to the different ways of spread spectrum, spread spectrum technology can be divided into direct sequence spread spectrum technology, frequency hopping technology and time hopping technology.

The direct sequence spread spectrum technology uses the spread spectrum code sequence with high code rate to expand the spectrum at the transmitting end, reduce the power spectrum density of the data link signal, make it difficult for the enemy to detect, so as to improve the anti-interference ability; the same spread spectrum code sequence is used at the receiving end to despread, and then after the spread spectrum code sequence modulation, the signal spectrum is widened, the power spectrum density is reduced, and the spread spectrum signal changes after demodulation For narrow-band signal, the power spectral density is increased and the system gain is increased several times, so the anti-interference ability is improved.

Frequency hopping technology uses the spread spectrum code sequence signal in the transmitter for frequency shift keying modulation, which makes the carrier frequency constantly jump, and uses the spread spectrum code sequence in the receiver to recover the signal. In the time domain, FH signal is a multi frequency FSK signal. In the frequency domain, FH signal will jump at unequal intervals in a wide frequency band. Compared with fixed frequency signal, as long as the enemy does not know the law of carrier frequency hopping, it is difficult to intercept our communication content. Even if some frequency points are interfered by the enemy, normal communication can still be carried out on other frequency points that have not been interfered, so it has good anti-interference ability.
The time hopping technology stores the input data in the transmitter first, and then the on-off switch is controlled by the spread spectrum code sequence of the spread spectrum code generator. After two-phase or four-phase modulation, it is transmitted after RF modulation. At the receiving end, the IF signal output by the RF receiver is controlled by the same spreading code sequence generated locally as the transmitting end, and then sent to the data memory by the two-phase or four phase demodulator. After timing, the output data can be recovered correctly as long as the two ends of the receiving and transmitting are in strict synchronization in time.

2.2.3.multiple input and multiple output technology. Multiple input and multiple output technology refers to the wireless communication technology that transmits signals through multiple antennas at the transmitting end and receives signals through multiple antennas at the receiving end. Multi input and multi output technology, combined with orthogonal frequency division multiplexing technology and space-time coding, can realize space diversity, time diversity and frequency diversity, and can realize anti-interference in the space, time and frequency domain [13]. At present, how to apply multi input and multi output technology to data link anti-jamming technology still needs to be studied on antenna configuration, power distribution, signal detection and other issues.

2.2.4. orthogonal frequency division multiplexing technology. Orthogonal frequency division multiplexing technology divides the channel into several orthogonal subchannels, transforms the high-speed data signals into parallel low-speed data streams, and then modulates them to the subchannels for transmission. Orthogonal signals can be separated in the receiver by using correlation technology to reduce the interference between sub channels. If the signal bandwidth on each subchannel is less than the correlation bandwidth of the signal, then each subchannel can be seen as flat fading, thus eliminating intersymbol interference [14].

2.2.5. adaptive antenna technology. At present, the data link of UAV mostly adopts the directional antenna technology for anti-jamming. The directional antenna means that the electromagnetic wave transmitted and received in one or several specific directions is very strong, while the electromagnetic wave received by the transmitter in other directions is 0 or very small. The purpose of anti-jamming is achieved by suppressing the received interference signal. The narrower the beam, the stronger the concealment and the stronger the anti-jamming ability.
Adaptive antenna technology uses the principle of phased array antenna to carry out spatial filtering for beams from all directions. It adjusts the excitation of antenna elements, optimizes the antenna array pattern, processes and identifies interference signals by using digital signal processing technology, forms the zero point of beam in the direction of dry interference source, suppresses the signals beyond the receiving direction, and reduces interference Reduce electromagnetic environmental pollution [15]. The disadvantage of adaptive antenna is to form a blind area in the direction of zero point, which affects the normal use of users in this area.

2.2.6.cognitive radio technology. Cognitive radio is an intelligent wireless communication system. Through cognitive and active learning of the surrounding radio environment, the working parameters of the communication system, such as coding mode, modulation mode, working frequency and transmission power, are changed in real time. The idle spectrum is dynamically detected and utilized to adapt to the changes of the external radio environment [8].

The UAV data link can use cognitive radio technology to monitor the interference in the communication frequency band. According to the characteristics of the interference source signal, it can change the channel coding mode, spread spectrum code sequence, frequency hopping pattern, time hopping mode, power configuration, antenna mode and other parameters of the data link in real time, reasonably use the radio spectrum resources, and improve the information transmission ability and anti-interference ability [16].

3. Key technical problems

At present, the challenges faced by UAV data link anti-jamming technology are as follows:

3.1. long distance transmission loss

Long distance transmission loss is the most serious challenge faced by UAV data link system. It includes the reduction of power attenuation and spectrum efficiency, transmission delay, increase of system error rate and packet loss rate, and decrease of receiver SNR.

3.2. Doppler shift of high speed motion

The high-speed motion of UAV will bring Doppler shift to the receiver, and the magnitude of Doppler shift is proportional to the speed of motion and inversely proportional to the wavelength. Because the frequency of electromagnetic wave is inversely proportional to the wavelength, the lower the working frequency band is, the smaller the Doppler frequency shift is for the UAV in high-speed motion state; however, the spectrum resource in the low-frequency band is very tight, and the Doppler frequency shift is a great challenge for the UAV data link.

3.3. signal fading caused by roadblocks

The main frequency band of UAV data link application is microwave (300MHz~3000ghz). The microwave link has higher available bandwidth, but the microwave frequency is high, the wavelength is very short, there is no diffraction function, the communication will cause shadow fading and multi-path fading.

3.4. artificial non malicious interference and malicious interference

Non malicious interference refers to the interference of the radio signals of other devices in the spectrum environment to the data link of UAV, while malicious interference mainly exists in the military field, mainly including partial pressure interference and deceptive interference.

3.5. spectrum environment congestion

Due to the shortage of spectrum resources, it is bound to be threatened by external interference signals. The traditional UAV data link frequency band includes HF, VHF and Satcom. Satcom frequency band can not guarantee that every data transmission phase can be used, but HF and VHF become more and more crowded. In China, the available frequency bands of UAV issued by the Ministry of industry and information technology only include 840.5mhz ~ 845mhz, 1430mhz ~ 1444mhz and 2408mhz ~ 2440mhz.

4. Future development trend

The future development trend of anti-jamming technology of UAV data link system is as follows:

(1) have a wide range of spectrum management and distribution capabilities, such as high throughput, high reliability physical layer and MAC layer protocols and strategies.

(2) the ability to solve the power attenuation of long-distance information transmission, such as low-power MIMO system integration.

(3) improve the ability of anti malicious interference and non malicious interference, such as spread spectrum, frequency hopping, multi-source cooperative communication, asynchronous cooperative information transmission, cognitive radio and other technologies.

IOP Conf. Series: Materials Science and Engineering **816** (2020) 012011 doi:10.1088/1757-899X/816/1/012011

5. Conclusion

With the application of UAV in various fields more and more widely, its data link system is facing the challenges of spectrum resource shortage, complex environment, vulnerable to environmental interference and human interference. Traditional spread spectrum communication technology can not meet the reliability requirements of UAV data link. Therefore, more flexible and Adaptive Anti-jamming Technology, such as intelligent sensing spectrum environment and system reconstruction, is needed to improve the anti-jamming ability of UAV and ensure the quality of communication link.

6. Reference

[1] Translated by Wu Hanping et al. Introduction to unmanned helicopter system [M]. Beijing: Electronic Industry Press, 2003

[2] Wen Fuzhong, Gou Xiaobo, Cheng Hao. EMC design of electronic warfare based on UAV [J]. Electronic information warfare technology, 2017, 32 (2): 32-37

[3] Shen Yan'an. Electromagnetic environment adaptability simulation of UAV remote control data link [J]. Value engineering. 2012 (5): 178-179

[4] Zhang Xiaodong, Chen Yaya, Xiao Xuerong. Research on electromagnetic interference mechanism of UAV main remote control data link [J]. Journal of microwave science, 2016, 32 (2): 90-96

[5] Ni Zhaoxu, Xue Xiaoliang, Wang Yang. Simulation and design of airborne antenna layout of air ground data link of unmanned helicopter [J]. Avionics technology. 2018, 49 (1): 7-12

[6] Yan Yunbin, Tian Qingmin, Wang Yongchuan, et al. Anti interference performance evaluation index and test method of UAV data link system [J]. Computer measurement and control, 2015, 23 (12): 3925-3928

[7] Liu Xuguang, Zhou Changzheng, Wang Manxi, et al. Modeling and analysis of anti-interference performance of multi band frequency modulation signal [J]. Military automation, 2015 (1): 56-59

[8] Liu Yiwei, park Haiyin, Xiao Lin, etc. overview of anti-jamming technology of UAV data link [J]. Aircraft design, 2017, 37 (6): 13-16, 21

[9] Zhang Xin. Deception interference of tactical data link [D]. Xi'an: Xi'an University of Electronic Science and technology. 2004

[10] Xia Linying. Research on tactical data link technology [D]. Xi'an: Northwest University of technology, 2007

[11] Yu Gefei, Wang Zhenhua, Li Ying, et al. Application of channel coding in UAV data link [J]. Journal of Chongqing University of Posts and Telecommunications: Natural Science Edition, 2014 (6): 845-849

[12] Rao Jun. spread spectrum code analysis of Spread Spectrum TT & C link of multiple UAVs [J]. Telecommunication technology, 2018, 58 (10): 73-78

[13] Luo Wencui. Research on UAV cooperation multi input multi output technology [D]. Xi'an: Xi'an University of Electronic Science and technology, 2017

[14] Huang wenzhun, Wang Yongsheng, ye Xiangyang. Research on a new anti-jamming technology of UAV data link [J]. Chinese Journal of Aeronautics, 2008 (2): 49-56

[15] Meng Lifeng, Han Yuhui. Application of smart antenna technology in UAV system [C]. National Symposium on remote sensing, telemetry and remote control, 2004

[16] Guo Caili, Zhang Tiankui, Zeng Zhimin, et al. Development and research status of cognitive radio technology at home and abroad [J]. Modern telecommunication technology, 2006 (6): 29-34

IOP Conf. Series: Materials Science and Engineering **816** (2020) 012012 doi:10.1088/1757-899X/816/1/012012

Effects of a Splitter and an Endplate on Flow-induced Vibration Characteristics of Cantilevered Rectangular and D-section Prisms

La Ode Ahmad Barata[1,2*], Takahiro Kiwata[3], Takaaki Kono[3] and Toshiyuki Ueno[4]

[1]Graduate School of Natural Science and Technology, Kanazawa University, Kakuma-machi, Kanazawa-shi, 920-1192, JAPAN
[2]Mechanical Eng. Department, Halu Oleo University, Kendari, 93231, Indonesia
[3]School of Mechanical Engineering, Kanazawa University, JAPAN
[4]School of Electrical, Information and Communication Engineering, Kanazawa University, JAPAN

*E-mail: ahmad.barata@uho.ac.id

Abstract The effects of a splitter and an end plate on the flow-induced vibration characteristics of cantilevered rectangular and D-section prisms with the different finite lengths (aspect ratios *L/H*) and depth ratios (side ratio *D/H* of 0.2 and 0.5) have been investigated in a water tunnel. In order to suppress the alternating vortex shedding in the wake of buff bodies, a splitter plate was installed behind the prisms with a small gap and sufficient length to prevent vortex shedding intrusion. Meanwhile, to maintain a two-dimensional flow structure, an endplate was attached to the prism tip end. Mounting the splitter plate enhanced the response amplitude for all prisms. In this case, interferences of amplitude signal spectra for the period of displacement vanished on the prism with a side ratio of 0.5 yielding a uniform waveform on response amplitude. The attached endplate does not take an important role in enhancing the dynamic response of the prisms with a side ratio of 0.5 but it suppressed the response amplitude of prisms with $D/H = 0.2$.

1. Introduction

The features of rectangular prisms with sharp edges in a uniform flow are the generation of alternate vortices behind the prism and the separation bubble on the side surfaces. Flow characteristics of fluid forces and vortex shedding frequency are dramatically changed by the side ratio of rectangular prism [1,2]. Because of these features, the elastically mounted rectangular prisms yield the flow-induced vibration in the cross-flow direction by the vortex-induced and the galloping vibrations. The dynamic response of prism depends on some parameters such as the system damping and after body shape (i.e. the structural part of a bluff body downstream of the flow separation points) [3]. In the case of finite rectangular prism mounted elastically in the water tunnel, Kiwata et al. [4] investigated the effect of side ratio on the response amplitude characteristics of rectangular and D-section prisms with the side ratios of less than $D/H = 0.5$, where D is the depth of a bluff body in the flow direction, and H is the height of a bluff body normal to the flow direction. It is found that the response amplitude of D-section prism with $D/H = 0.23$ vibrated at lower reduced velocity than the rectangular prism with $D/H = 0.2$, and the increment rate of response amplitude of D-section prism with $D/H = 0.23$ were higher than the other prisms.

The flow features of prisms with low aspect ratios of $L/H \leq 5$, where L is the span length of a prism, are considerably influenced by three-dimensional vortex structure from the tip of a prism and the alternating vortex shedding behind the bluff body [5]. If the endplate is installed on the tip of a prism, the three-dimensional flow structure of wake can be modified to be the two-dimensional flow structure of wake. The wake structure and the galloping properties for both attached and unattached endplate conditions can be found in the references [6,7,8]. Meanwhile, the suppression of alternating vortex shedding behind the prism around the critical side ratio was carried out by the installation of a splitter plate as references [9,10,11].

To develop the power generation system using the iron-gallium alloy and the flow-induced vibration, we focus on the low-speed galloping for the cantilever mounted rectangular and D-section prisms with side ratios of $D/H \leq 0.5$. Our main objective is to increase the response amplitude of prisms by installing a splitter and an endplate. The effects of a splitter and an endplate on the transverse vibration characteristics of the cantilevered rectangular and D-section prisms was investigated experimentally in a water tunnel.

2. Experimental Approaches

2.1 Experimental apparatus

The experiment of free vibration of cantilevered prisms was carried out in a water tunnel facility [4]. Figure 1 shows a schematic diagram of the test section in the water tunnel and the measurement instruments. The rectangular test section of water tunnel had a height of 400 mm, a width of 167 mm, and a length of 780 mm. The uniform flow velocity U was changed from 0.74 m/s to 2.7 m/s by controlling the pump's rotational speed and opening of the control valve and measured by a pitot tube and a digital differential pressure gauge (NAGANO KEIKE, GC50). A splitter plate with 200 mm (= $10H$) in length and 1.0 mm in thickness was installed behind the prism with a gap of about 6 mm. The splitter plate length is sufficient to suppress the interaction between the upper and lower sides of vortices shedding behind the prism [11]. Meanwhile, an endplate was attached at the tip of prisms. The endplate had a diameter of 60 mm (= $6H$) and a thickness of 1.0 mm.

2.2 Test models and experimental methods

As shown in figure 1, the prism was mounted elastically to a plate spring made from stainless steel and attached to the case wall on the test section. The prism was only vibrated transversely to the flow direction. Table 1 shows the specifications of the test models with and without an endplate. The

Table 1. The Specification of test models

Cross section	Rectangular						D-section					
	Side ratios (D/H)						Side ratios (D/H)					
	0.2			0.5			0.23			0.5		
L/H	f_c	δ	C_n	f_c	δ	C_n	f_c	δ	C_n	f_c	δ	C_n
Attached endplate												
2.5	31.9	0.051	0.91	24.1	0.047	0.84				27.34	0.072	1.26
5	32.9	0.072	1.26	22.8	0.062	1.10				25.63	0.079	1.40
7.5	23.0	0.038	0.66	15.5	0.055	0.97						0.00
10	24.8	0.044	0.78	16.4	0.026	0.45				17.72	0.070	1.23
Free end model												
2.5	38.5	0.022	0.39	26.4	0.023	0.40				29.79	0.013	0.24
5	38.7	0.026	0.46	24.8	0.026	0.46				28.08	0.014	0.25
7.5	26.4	0.050	0.89	16.5	0.030	0.52						0.00
10	26.5	0.026	0.46	16.4	0.031	0.55				19.04	0.011	0.20

rectangular and D-section prisms had a cross-section height H of 20 mm. As shown in figure 2, the side ratio D/H (D: depth of the prism in the flow direction) of the rectangular prism was 0.2 and 0.5. The side ratio D/H of the D-section prism was 0.23 and 0.5. The span lengths for both prisms L were varied between 2.5, 5.0, 7.5 and 10. The characteristic frequency f_c of the prism was adjusted to a constant value between 16 and 39 Hz by using the plate spring with different thickness and analyzed using FFT analyzer (ONO SOKKI, CF-5201). The reduced mass-damping parameter C_n ($=2\delta\rho_{st}/\rho_f$) was measured considering the initial displacement which is obtained by hitting the prism with a hammer in the still water, where ρ_{st}, δ, and ρ_f is the density of prism, the logarithmic decay rate of the structural damping parameter of a prism, and the water density respectively. Tip displacement y was measured using an acceleration sensor (Showa Measuring Instrument, 2302CW) implanted inside the tip of the test model and connected to the integrator (RION, UV-12 and UV-05). The signals output of the integrator was converted using 12-bit A/D converter with the sampling frequency of 2 kHz, for which 16,384 data points were recorded. The reduced velocity V_r ($= U/f_cH$) was varied from 1.3 to 5.0 by controlling the water velocity U.

3. Results and Discussions

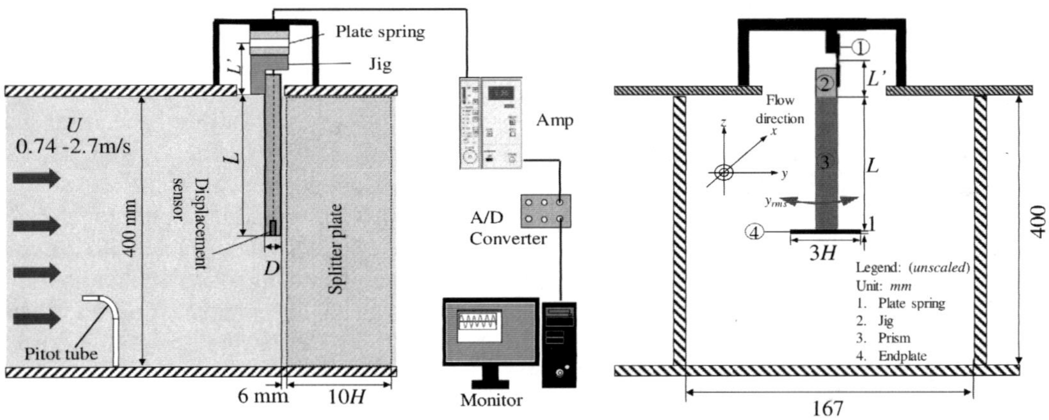

(a) Installed splitter plate at test section (Side view of the test section)
(b) Attached endplate at the tip of model (Front view of the test section)

Figure 1. Schematic diagram of the test section of water tunnel and the measurement instruments

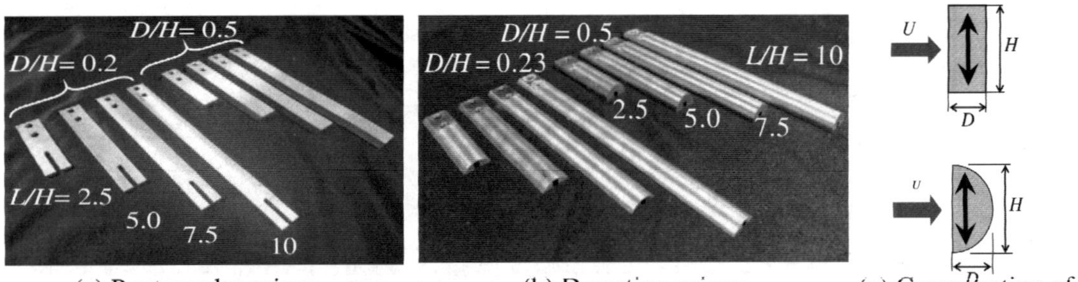

(a) Rectangular prisms
(b) D-section prisms
(c) Cross section of test models

Figure 2. Overview and cross section of the test models

3.1 The Effect of a splitter plate on response amplitude

The rectangular and D-section prisms with an aspect ratio of L/H = 2.5 has non-uniform amplitude, small amplitude, and low increment rate of the response amplitude. On the contrary, the prism with L/H = 10 has stable vibration and high increment rate of response amplitude as similar to the study in reference [3]. Hence, the prisms with aspect ratio L/H of 10 were a more suitable cross-section for harvesting energy from the transverse galloping vibration. Meanwhile, this paper focus on the effect of a splitter plate on transverse vibration characteristics of cantilevered rectangular and D-section with different aspect ratios of L/H =2.5 to 10 to enhance the dynamic response of those prisms.

3.1.1 Rectangular prisms. The response amplitudes of rectangular prisms with side ratios of D/H = 0.2 and 0.5, and aspect ratios of L/H = 2.5, 5.0, 7.5 and 10 are shown in figure 3. The vibration onset of the prism without a splitter plate was almost the same value as that with a splitter plate. For the aspect ratios of $L/H \geq 5.0$, the increment rate of the non-dimensional response amplitude η_{rms} after the vibration onset was almost the same value as that for the rectangular prisms with a side ratio of D/H = 0.2 and 0.5. For a side ratio of D/H=0.5 with a short span length of the prism L/H=2.5, the response amplitude of prism with a splitter plate increases more than that without a splitter plate. For the small side ratio of D/H=0.2 with a short span length of L/H=2.5, we did not find a meaningful improvement in response amplitude by inserting a splitter plate. However, the improvement in response amplitude was found in the prism with a side ratio of 0.5.

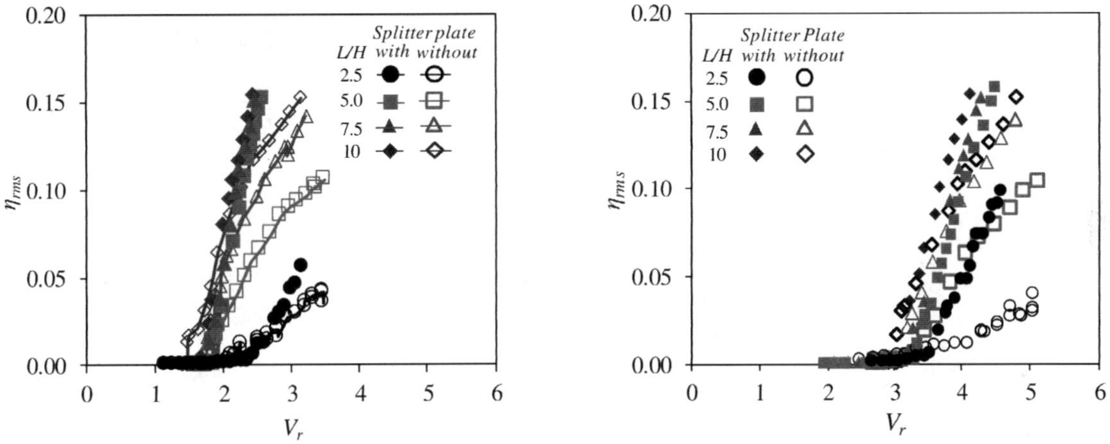

(a) D/H = 0.2 (b) D/H = 0.5

Figure 3. Response amplitudes of rectangular prisms without and with a splitter plate

Time histories and frequency spectra of the tip displacement for a rectangular prism with D/H = 0.5 are shown in figures 4 and 5. The rectangular and D-section prisms vibrated at the characteristic frequency f_c defined in table 1. In the case of a rectangular prism with D/H = 0.5, without a splitter plate, the amplitude of the vibration is not stable. As shown in figure 4(a), the rectangular prism with D/H = 0.5 and L/H = 10 has two components of the characteristic frequency i.e. prism's amplitude and the vortex shedding frequency. This phenomenon is causally related to the stationary rectangular prism with D/H = 0.5 in which it appears intermittently because of two different flow patterns yielding abruptly changes on pressure base suction [12]. However, when the splitter plate is inserted behind the prism, the vibration has uniform amplitude for each aspect ratio and those intermittent waveforms vanish for all aspect ratio and result in one sharp peak only on the FFT spectrum as shown in figure 4(b).

If the splitter plate was inserted behind a circular cylinder, the effect of vortex shedding interaction on the vortex-induced transverse vibration does not occur due to the suppression of the alternative vortex shedding. On the other hand, if the splitter plate was inserted behind a rectangular cylinder of $D/H \leq 0.5$, the low-speed galloping vibration occurs and instability increased. This is because the galloping instability does not relate to the frequency of the alternative vortex shedding, and it is closely related to the minus sign of the lift coefficient slop versus the attack of angle [13]. Vanishing a second peak adjacent to the sharp peak may represent increasing of the instability of galloping, and alter the critical point of this bluff bodies as mentioned in reference [14].

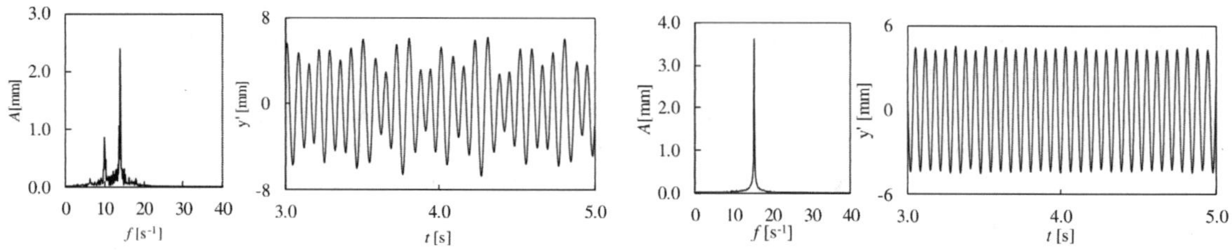

(a) Without splitter plate ($V_r = 4.95$) (b) With splitter plate ($V_r = 4.13$)

Figure 4. Time histories and frequency spectra of the tip displacement for a rectangular prism with $D/H = 0.5$ and $L/H = 10$

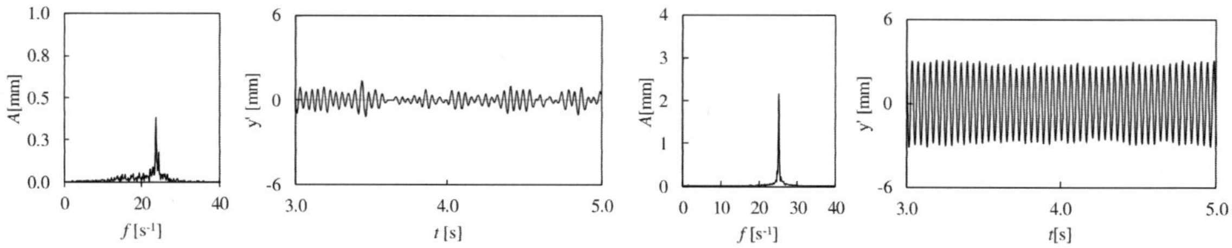

(a) Without splitter plate ($V_r = 5.04$) (b) With splitter plate ($V_r = 4.57$)

Figure 5. Time histories and frequency spectra of the tip displacement for a rectangular prism with $D/H = 0.5$ and $L/H = 2.5$

3.1.2 D-section prisms. The response amplitudes of D-section prisms with a side ratio of $D/H = 0.5$, and aspect ratios of $L/H = 2.5$, 5.0 and 10 are shown in figure 6. Inserting a splitter plate behind the D-section prism brings about high response amplitude response. In particular, the response amplitude of prisms with small aspect ratio of $L/H \leq 5.0$ was built up. Figure 7 shows time histories and frequency spectra of the tip displacement for a D-section prism with $D/H = 0.5$ and $L/H = 2.5$ without and with a splitter

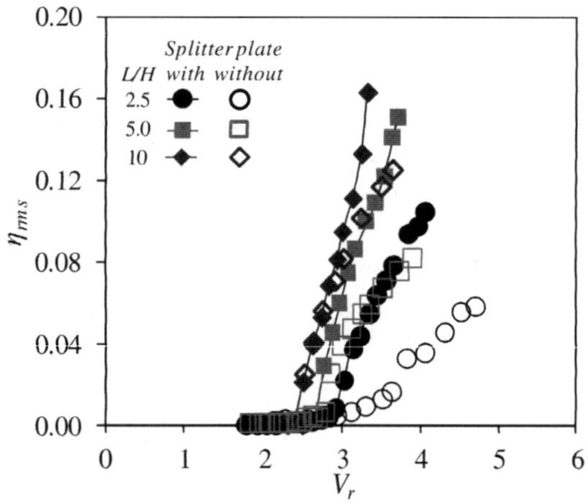

Figure 6. Response amplitudes of a D-section prism without and with a splitter plate for $D/H = 0.5$

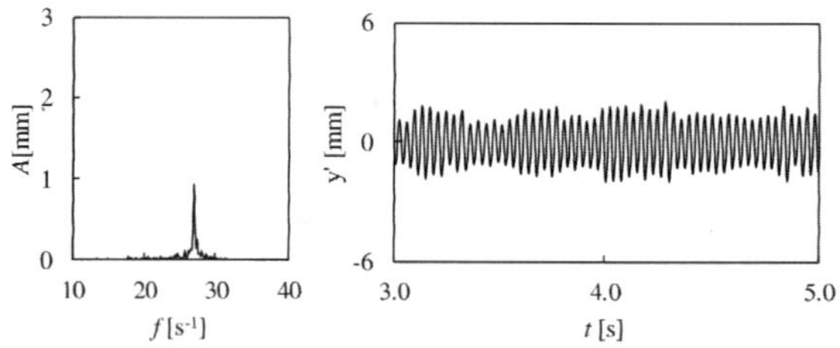

(a) Without splitter plate ($V_r = 4.70$)

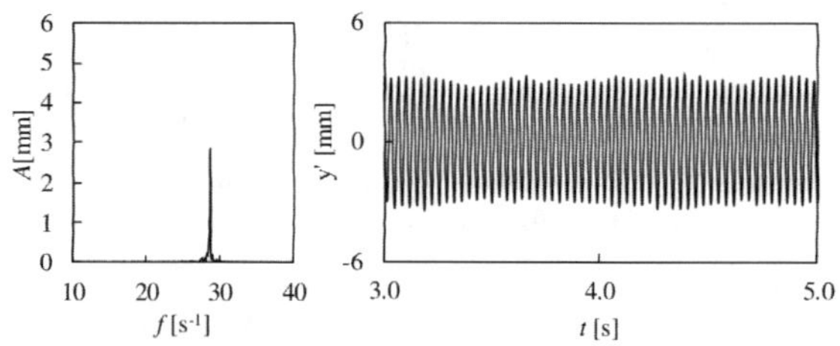

(b) With splitter plate ($V_r = 3.34$)

Figure 7. Time histories and frequency spectra of the tip displacement for a D-section prism with $D/H = 0.5$ and $L/H = 2.5$

plate. The FFT spectrum of a D-section with *D/H* = 0.5 has one sharp peak only. It means that the galloping behavior of D-section prisms is purely instability type and free from alternating vortices interferences at the after body. This is distinctly different from the rectangular prism at a similar side ratio. The effect of free end on the vibration characteristics is also observed like the rectangular prism that the D-section prism with a small aspect ratio of *L/H* = 2.5 without a splitter plate exhibits a slightly unstable response amplitude though only one dominant peak on the FFT spectrum (figure 7(a)). By inserting a splitter plate, the response amplitude becomes stable and increases.

3.2 Effect of an endplate on response amplitude

As shown in table 1, the characteristic frequency of prism with an endplate decreases about 2 Hz due to the mass of endplate. The reduced mass-damping parameter C_n of cantilevered rectangular and D-section prisms with attached an endplate also increases more than that without one. However, the endplate will be expected to diminish the swirling up vortex from the tip of prism [6]. This section focuses on the effect of an attached endplate on transverse vibration characteristics of cantilevered rectangular and D-section with different aspect ratios of *L/H* =2.5 to 10.

3.2.1 Rectangular prisms. The response amplitudes of rectangular prisms with side ratios of *D/H* = 0.2 and 0.5, and aspect ratios of *L/H* = 2.5, 5.0, 7.5 and 10 are shown in figure 8. The vibration onset of the prism without an endplate is almost the same value as that with an endplate. For an aspect ratio of *L/H* = 10, the non-dimensional response amplitude η_{rms} of the prism with attached endplate increases at the large value of reduced velocity than that without an endplate. For the rectangular prisms with *D/H* = 0.2 and *L/H* ≤ 5, the η_{rms} of the prism with attached endplate decreases than that without an endplate. The vibration of a rectangular prism with attached endplate for *L/H* = 2.5 suppress hardly while the prism with *L/H* > 5, the effect of attached endplate on the response amplitude is out of our prediction with any meaningless improvement on the response amplitude.
Time histories of the tip displacement for a rectangular prism with *D/H* = 0.2 and *L/H* = 2.5 and with an endplate are shown in figure 9. The amplitude of the vibration is not stable. As shown in figure 8(b), for the rectangular prisms with *D/H* = 0.5 and *L/H* ≤ 5, the η_{rms} of the prism with endplate is the same as that without an endplate in which attached endplate does not affect the amplitude of the prisms. Figure 10 shows the frequency spectra of the tip displacement at maximum response amplitude for a rectangular prism with *D/H* = 0.5 with the attached endplate. However, the rectangular prisms with *L/H* = 2.5 and 5.0 have one predominant spectrum of the characteristic frequency of the prism. This is in contrast with those without endplate which has two sharp peaks on the FFT spectra. By increasing the aspect ratio (*L/H*=10), two components of sharp peaks i.e. the characteristic frequency of the prism and the alternating vortex shedding frequency appear. The cause of increment of response amplitude (figure

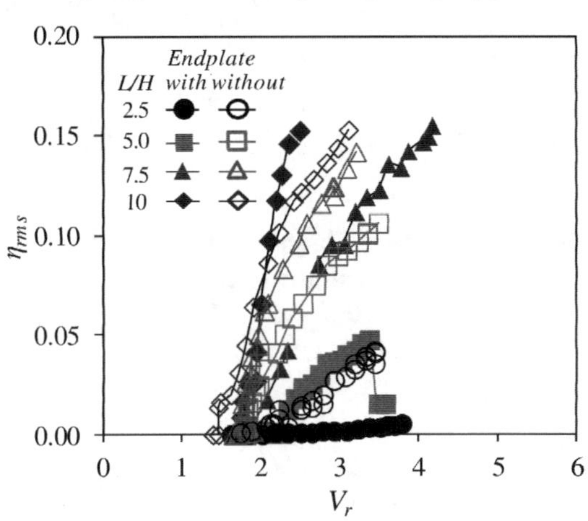

(a) *D/H* = 0.2

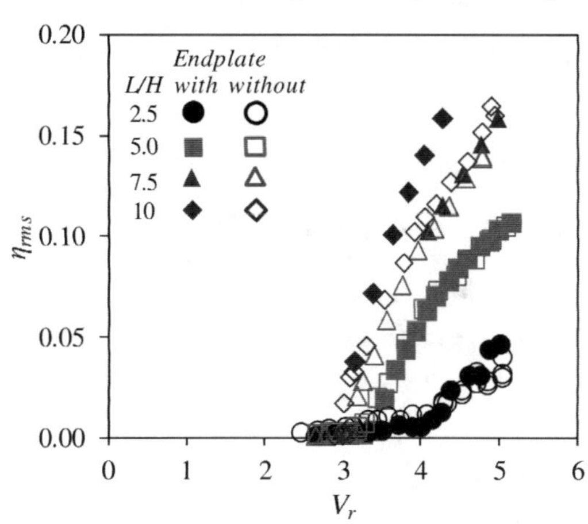

(b) *D/H* = 0.5

Figure 8. Response amplitudes of rectangular prisms without and with an endplate

8(b)) for L/H = 10 may be attributed to the alternating vortex shedding frequency. We have to confirm the interference of wake fluctuation on the response amplitude of the prism with a large aspect ratio (L/H = 10) in further study.

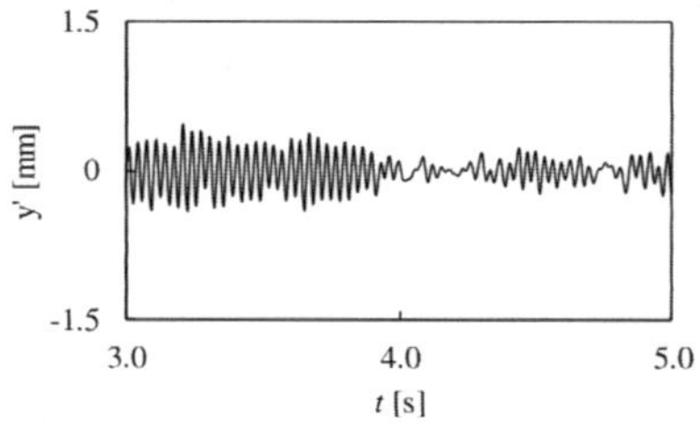

Figure 9. Time histories of the tip displacement for a rectangular prism with D/H = 0.2 and L/H = 2.5 and with an end plate (V_r = 3.78)

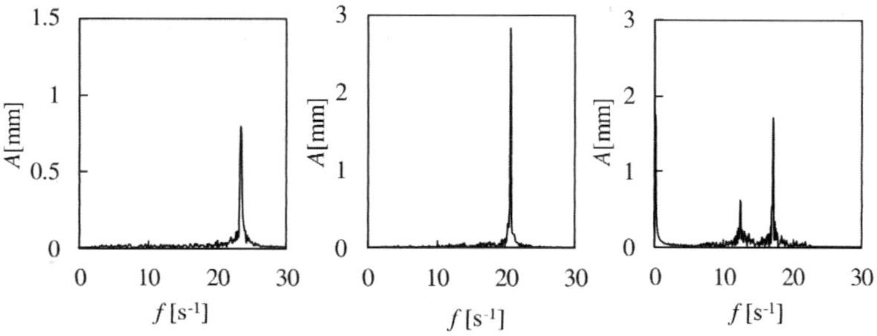

(a) L/H = 2.5 (V_r = 5.01)
(b) L/H = 5.0 (V_r = 5.14)
(c) L/H = 10 (V_r =4.56)

Figure 10. Frequency spectra of the tip displacement for rectangular prisms with end plate at maximum response amplitude (D/H = 0.5)

3.2.2 D-section Prisms. Figure 11 shows the response amplitudes of D-section prism with a side ratio of D/H = 0.5, and aspect ratios of L/H = 2.5, 5.0 and 10. The non-dimensional response amplitude η_{rms} of the prism with endplate increases slightly at the large value of reduced velocity rather than that without an endplate. Figure 12 shows the time histories and frequency spectra of the tip displacement for a D-section prism with D/H = 0.5 and L/H = 2.5, and with an attached endplate. The frequency spectrum has one predominate component of the characteristic frequency of the prism which is different from the rectangular prism in the similar side ratio. Okajima et al (2004) described that the endplate effect prolonged reduced resonant velocity range in case of the circular cylinder by resulting in two excitation regions. The steeper graph of response amplitude in figure 11 may be attributable to this phenomenon.

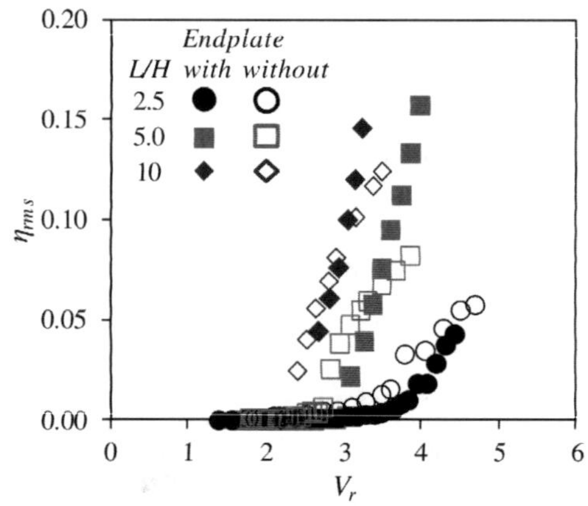

Figure 11. Response amplitudes of the D-section prisms with D/H=0.5 without and with attached endplate

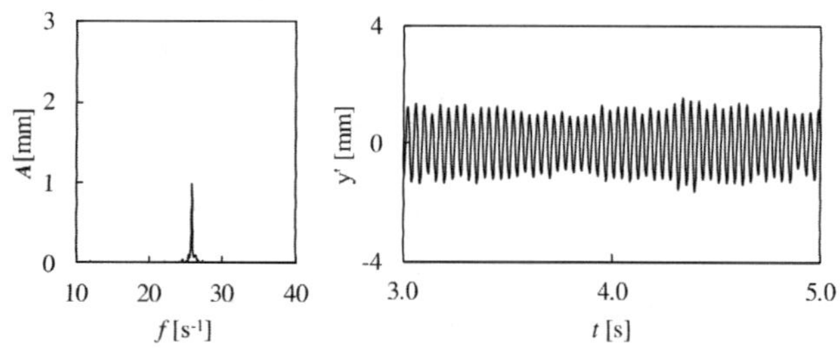

Figure 12. Time histories and frequency spectra of the tip displacement for a D-section prism with D/H = 0.5 and L/H = 2.5 and with an attached endplate (V_r =4.42)

4. Conclusions

Free-vibration tests involving a cantilevered rectangular prism and a D-section prism with a side ratio of less than 0.5 were performed in a water tunnel. The effects of a splitter and an endplate on flow-induced vibration characteristics of prisms were investigated. However, further investigation about flow wake structure behavior in presence of splitter plate and endplate for slender prism is curious for future research. The recent conclusions of the present study are as follows:

1. Inserting a splitter plate behind prism improve the response amplitudes for finite cantilevered rectangular and D-section prisms by stabilizing response.
2. Attaching an endplate on the tip of prisms does not affect the response amplitude for prisms with a side ratio of $D/H = 0.5$ and aspect ratios of $L/H \leq 5.0$. However, attached endplate can suppress the response amplitude for the rectangular prism with a side ratio of $D/H = 0.2$ and the aspect ratios of $L/H \leq 5.0$.

5. References

[1] A. Okajima. Strouhal numbers of rectangular cylinders. *J. Fluid Mech.*. 123(1), (1982), pp. 379-398.
[2] Y. Nakamura and T. Matsukawa. Vortex excitation of rectangular cylinders with a long side normal to the flow. *J. Fluid Mech.*, Vol. 180, No. 1,(1987), pp. 171.-191.
[3] Y. Nakamura and K. Hirata. Pressure fluctuations on oscillating rectangular cylinders with the long side normal to the flow. *J. Fluids Struct.*, Vol. 5, No.. 2, (1991), pp. 165–183.
[4] T. Kiwata, M. Yamaguchi, T. Kono, and T. Ueno. Water tunnel experiments on transverse-galloping of cantilevered rectangular and D-section prisms. *J. Fluid Sci. Technol.*, Vol. 9, No. 3,(2014), pp. 1–5
[5] D. Sumner, Flow above the free end of a surface-mounted finite-height circular cylinder: A review. *J. Fluids Struct.*, Vol. 43, (2013), pp. 41–63.
[6] T. L. Morse, R. N. Govardhan, and C. H. K. Williamson. The effect of end conditions on the vortex-induced vibration of cylinders. *J. Fluids Struct.*, Vol. 24, No. 8, (2008), pp. 1227–1239.
[7] T. A. Fox and G. S. West. On the use of end plates with circular cylinders. *Exp. Fluids*, Vol. 9, No. 4, (1990), pp. 237–239.
[8] A. Slaouti and J. H. Gerrard. An experimental investigation of the end effects on the wake of a circular cylinder towed through water at low Reynolds numbers. *J. Fluid Mech.*, Vol. 112, No.1, (1981), pp. 297–314.
[9] Y. Nakamura, K. Hirata, and T. Urabe. Galloping of rectangular cylinders in the presence of a splitter plate. *J. Fluids Struct.*, Vol. 5, No.5, (1991), pp. 521–549.
[10] A. R. Ogunremi and D. Sumner. The effect of a splitter plate on the flow around a finite prism. *J. Fluids Struct.*, Vol. 59, (2015), pp. 1–21.
[11] M. K. Chauhan, S. Dutta, B. S. More, and B. K. Gandhi. Experimental investigation of flow over a square cylinder with an attached splitter plate at intermediate Reynolds number. *J. Fluids Struct.*, Vol. 76,(2018), pp. 319–335.
[12] Y. Ohya. Note on a Discontinuous Change in Wake Pattern for a Rectangular Cylinder. *J. Fluids Struct.*, Vol. 8, Issue 3, (1994), pp. 325-330.
[13] G. Alonso, J. Meseguer. A parametric study of the galloping stability of two-dimensional triangular cross-section bodies. *J. Wind Eng. Ind. Aerod.*, Vol. 94, (2006), pp.241–253.
[14] Y. Nakamura and K. Hirata. Critical geometry of oscillating bluff bodies. *J. Fluid Mech.*, Vol. 208, No.1, (1989), pp. 375–393.
[15] Okajima et al. Flow-induced in-line oscillation of a circular cylinder. *European Journal of Fluid Mechanics B/Fluid* 23 (2004), pp 115-125

Acknowledgments
This research is a part of work on Ph.D. research fellowship between Kanazawa University and Directorate General of Higher Education of Indonesia through KU-DIKTI program funded by Indonesia Endowment Fund for Education (LPDP). The author also would like to thank Mr. Keita Kawakami and Mr. Tomohiro Kuratani for their help during this experiment.

IOP Conf. Series: Materials Science and Engineering **816** (2020) 012013 doi:10.1088/1757-899X/816/1/012013

Research on Anti-Vibration based on Screw Fixation and Friction-Assisted Fixation Method for the Space Payloads

F F Wang[1], L Zhang[1], Z H Qiao[1], L Li[2] and D K Wang[3]

[1] Key Laboratory of Space Utilization, Technology and Engineering Center for space Utilization, Chinese Academy of Sciences, Beijing, 100094, China
[2] Qian Xuesen Laboratory of Space Technology, Beijing, 100094, China
[3] Beijing Institute of Electronic System Engineering, Beijing, 100854, China

wangff16@csu.ac.cn

Abstract. Rack is kind of the experimental platform that used in the space station for performing scientific experiment. Many payloads are mounted on Rack with limited operating space. The shipment of a Rack to its final destination in orbit is far more complicated than for all other commercial products. The launch vehicle-induced vibration and shock environment can be a payload killer. Based on the limited operating space, this paper proposes a screw fixation and friction-assisted fixation method. Through establishing the force and the parameters of the screw fixation and friction-assisted fixation based on the payload, deriving the mathematical model and carrying out the dynamic environment test. Test results match theoretical results quite well. The screw fixation and friction-assisted fixation method can effectively resist excessive dynamic and shock loads.

1. Introduction

As the development of Chinese space technology, many space loads with various mission are investigated a lot. Rack is kind of the experimental platform that used in the space station for performing scientific experiment. Like satellites, the Rack undergoes design, fabrication, test, and shipment. However, the shipment of a Rack to its final destination in orbit is far more complicated than for all other commercial products. Excessive dynamic and shock loads can be a satellite killer causing permanent damage to electronics, optics, and other sensitive equipment. To compensate for the harsh dynamic environment, payloads must be designed and tested to very high dynamic levels [1~5].

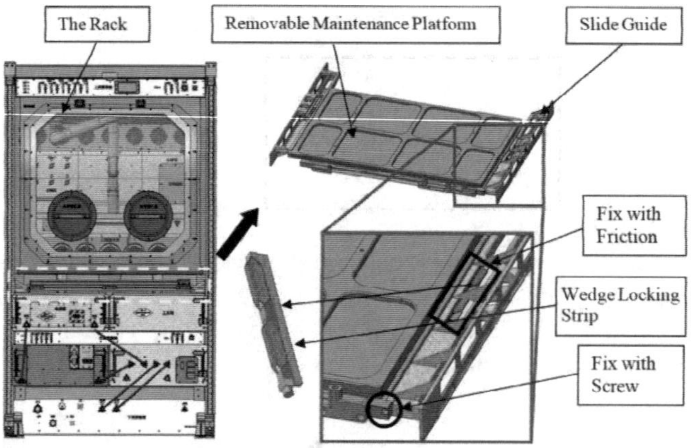

Figure 1. Removable maintenance platform and installation method

In this paper, an installation method is introduced to fix the payload (removable maintenance platform) on the rack with limited operating space. Through screw fixation and friction-assisted fixation method, as shown in figure 1, the safety and reliability of removable maintenance platform during the launching process can be guaranteed.

2. Model and Theoretical Analysis

2.1. Model

The parameterized model and the force of the removable maintenance platform are shown in figure 2, and figure 2a) is the top view, figure 2b) is the side view. The force of the wedge locking strip is shown in figure 3. And the force of the friction-assisted fixation is shown in figure 4.

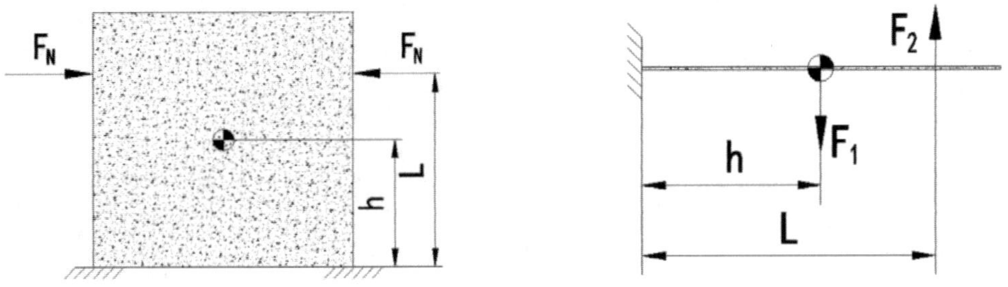

a) the top view b) the sideview

Figure 2. The parameterized model and the force of the removable maintenance platform

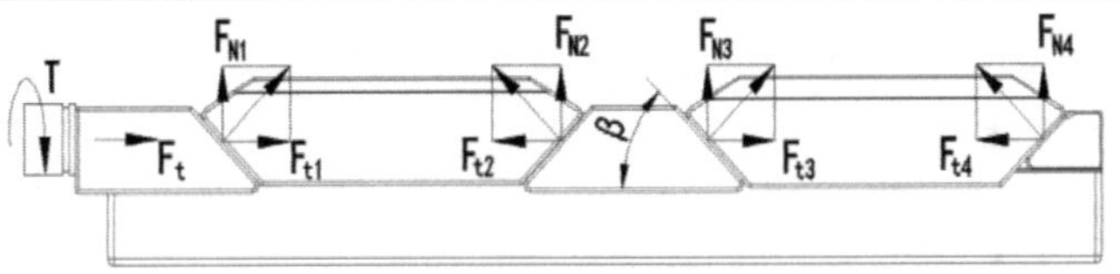

Figure 3. The force of the wedge locking strip

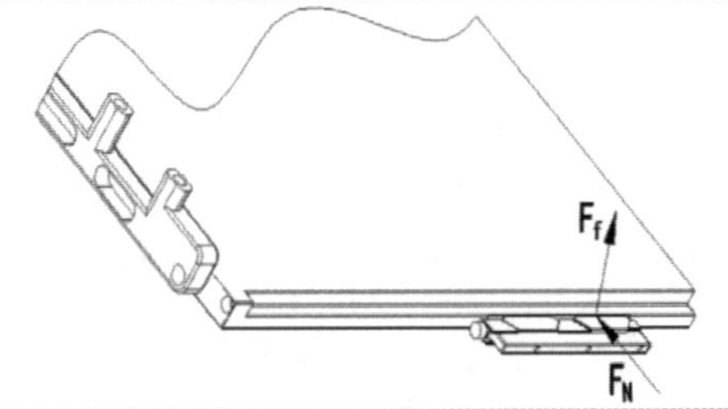

Figure 4. The force of the friction-assisted fixation

2.2. Force Analysis

According to the figure 2, we can get the impact force of removable maintenance platform

$$F_1 = ma \tag{1}$$

where, m is the mass of removable maintenance platform; a is the impact acceleration.
And the following force balance equation can be obtained

$$F_1 \times h = F_2 \times L \tag{2}$$

where, L is the distance between the friction force and the fixed point; h is the distance between the center of mass and the fixed point; F_2 is the force which can withstand the impact load.
So

$$F_2 = \frac{mah}{L} \tag{3}$$

According to the Fig. 3, we can get the force produced by the torque applied to the screw

$$F_t = \frac{T}{D} \tag{4}$$

where, T is the torque applied to the screw; D is the diameter of the screw.
And the following equations can be obtained

$$\begin{cases} F_{N1} = F_{t1} \cot\beta \\ F_{N2} = F_{t2} \cot\beta \\ F_{N3} = F_{t3} \cot\beta \\ F_{N4} = F_{t4} \cot\beta \\ F_{t1} = F_{t2} = F_{t3} = F_{t4} = F_t \end{cases} \tag{5}$$

where, β is the angle of wedge block; F_{N1}, F_{N2}, F_{N3}, F_{N4} are the pressure component of the wedge block at different positions on the removable maintenance platform.

According to figure 4, the force of friction generated by a single wedge locking strip can be obtained

$$\begin{cases} F_f = \mu F_N \\ F_N = F_{N1} + F_{N2} + F_{N3} + F_{N4} \end{cases} \tag{6}$$

So

$$F_f = \frac{4\mu T \cot\beta}{D} \tag{7}$$

According to the current design, the values of the above variables are shown in the table 1.

Table 1. The values of the different variables

Variables	m(kg)	a(m/s^2)	h(m)	L(m)	μ	T (N.m)	β(°)	D(m)
Value	12	300	0.25	0.4	0.3	7	45	6×10^{-3}

According to the above formula and table, we can get

$$\begin{cases} F_2 = 2250\text{N} \\ F_f = 1400\text{N} \end{cases} \tag{8}$$

So

$$2F_f > F_2 \tag{9}$$

Therefore, the design can meet the requirements of safety and reliability.

3. Fabrication and Assembling

The wedge locking strip is manufactured and assembled as shown in figure 5. In order to increase the friction force, the friction contact part is knurled.

Removable maintenance platform and slider guides are manufactured and assembled, which integration with the Rack, as shown in figure 6.

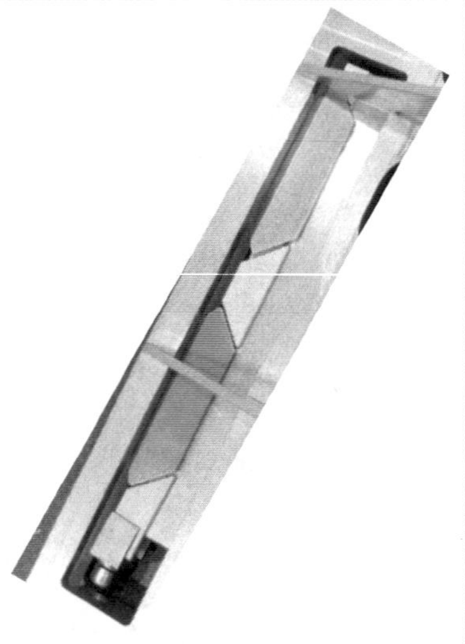

Figure 5. The real product of the wedge locking strip

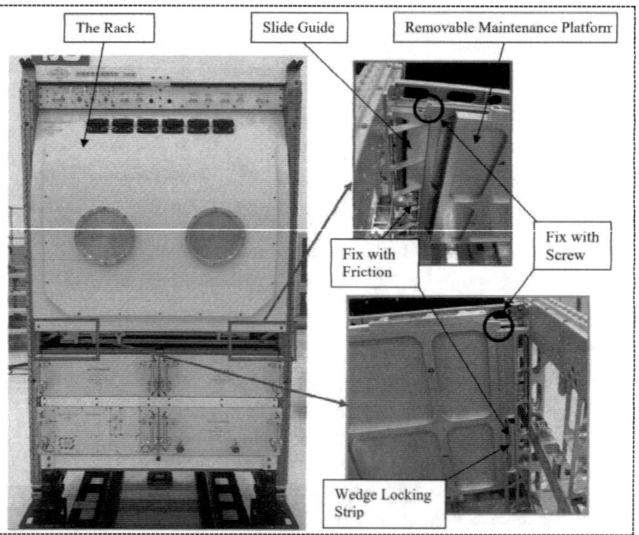

Figure 6. Removable maintenance platform integration with the Rack

4. Test

The acceleration measurement points are pasted in the corresponding position of removable maintenance platform as shown in figure 7.

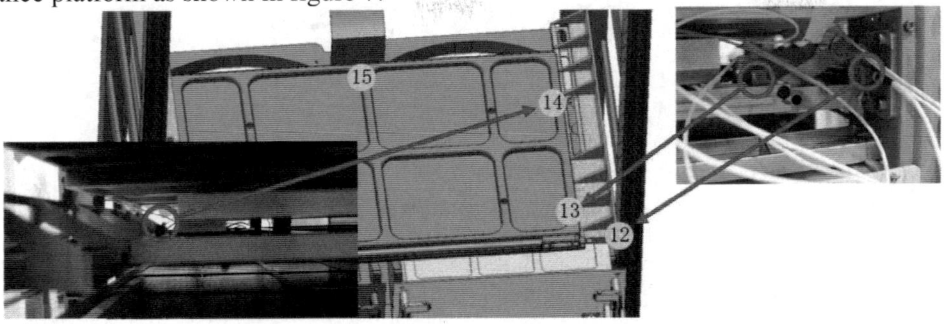

Figure 7. The acceleration measurement points of the removable maintenance platform

The dynamic environment testing was carried out on the Rack, as shown in figure 8. And the conditions of the dynamic environment testing conditions are shown in table 2.

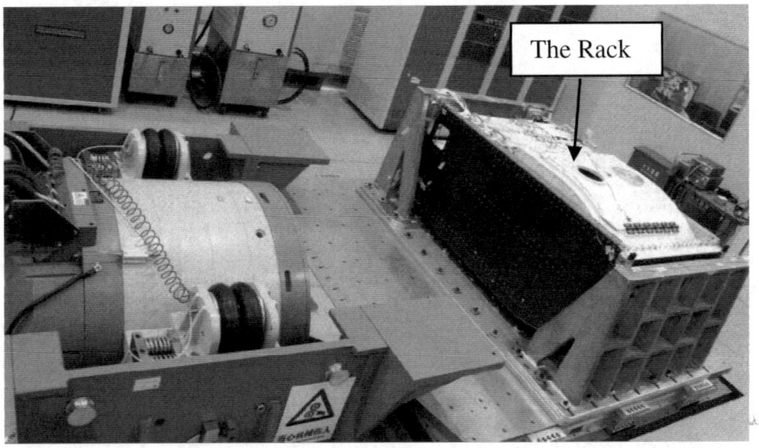

Figure 8. The dynamic environment testing platform

Table 2. The test condition of sinusoidal vibration

Parameter Name	Parameter Values			
	Frequency Domain（Hz）			
	4-10	10-17	17-60	60-100
Value of Amplitude	7.5mm	3g	5g	3g
Sweep Frequency	2 oct/min			
Loading Direction	Three axial directions			

The mechanical conditions shown in table 2 are applied to the Rack. The acceleration curve of the acceleration measurement points as shown in figure 7 can be obtained, as shown in figure 9. It can be seen from figure 9 that the maximum acceleration is 300 m/s^2 at the acceleration measurement point 15.

After the above test conditions, the output curve at 15 measurement points before and after the 0.2g characteristic scanning can be obtained, as shown in figure 10.

The above test results show that the frequency sweep results before and after the test are consistent. The installation state has not changed to meet the design requirements.

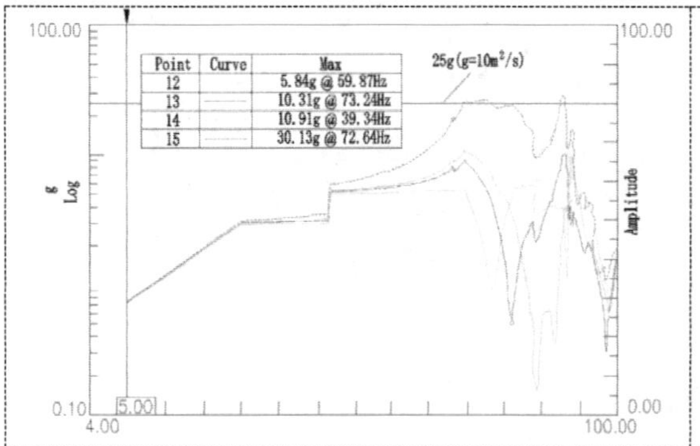

Figure 9. The acceleration curve of the different measurement points

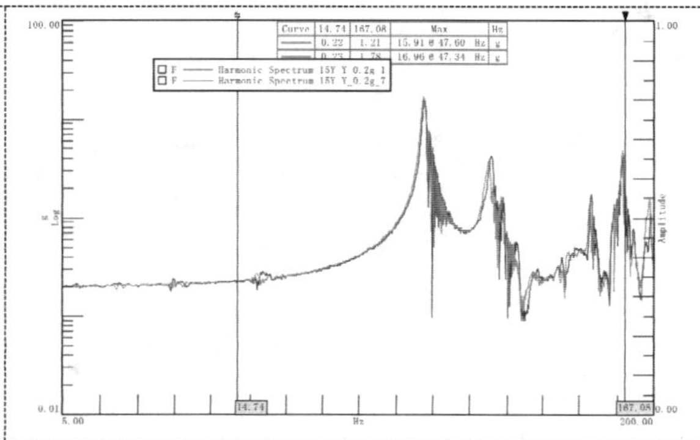

Figure 10. The output curves of the point 15 before and after the 0.2g characteristic scanning

5. Conclusions

In this paper, the screw fixation and friction-assisted fixation Method has been proposed. The mathematical model was established and the dynamic environment test were carried out. Test results match theoretical results quite well. The results show that the screw fixation and friction-assisted fixation method can ensure the safety and reliability of the payload.

6. References

[1] Johnson C D, Wilke P S, and Pendleton S C. SoftRide Vibration and Shock Isolation Systems that Protect Spacecraft from Launch Dynamic Environments, Proceedings of the 38th Aerospace Mechanisms Symposium, Langley Research Center, 17-19, May. (2006)

[2] Wang S K, Wang K, and Zhou Y L. Development of the Varying Gravity Rack (VGR) for the Chinese Space Station. Microgravity Science and Technology, 31:95-107(2019)

[3] Roques, R.: BIOLAB: Performance Assessment of Flight, Ground and Science Models. In: 54th International Astronautical Congress of the International Astronautical Federation, the International Academy of Astronautics, and the International Institute of Space Law. International Astronautical Congress (IAF). American Institute of Aeronautics and Astronautics, (2003)

[4] Uffmann, L., Hambloch, P., Diefenbach, A., Willnecker, R., Steinbach, S., Enz, T., Simicic, D., Nemirovski, A., Brisset, J., Weiβ, P., Aicher, W.: Decentralized payload operations: ESA' smaterialssciencelaboratoryinNASA' smaterialsscience research rack. In: Proc. of the SpaceOps 2010 Conference, Huntsville, Alabama, AIAA 2010–1945 (2010)

[5] Hu, W., Long, M., Kang, Q., Xie, J., Hou, M., Zhao, J., Duan, L., Wang, S.: Space experimental studies of microgravity fluid science in China. Chin. Sci. Bull. 54(22), 4035–4048 (2009)

Acknowledgments

We want to thank Shenyang Institute of Automation, Chinese Academy of Sciences for providing the dynamic environment test results, and acknowledge support from the national project (Y7140211XN).

IOP Conf. Series: Materials Science and Engineering **816** (2020) 012014 doi:10.1088/1757-899X/816/1/012014

A Novel Hand Pallet Truck and Multi-objective Optimization Design of It's Lifting Mechanism

Bo Yang, Jianhui Yuan*, Min Chen and Fenhua Yu

Unit 96815 of PLA, Jingdezhen, 333000, China

*Corresponding author's e-mail: tech_centre@163.com

Abstract. Hand pallet truck is widely used to carry baggage and cargoes in daily life and work. However, rotary-lift based luggage hand pallet truck is inconvenient to handle and difficult to carry heavy goods. Hydraulic cylinder based hand pallet truck is also used commonly, but its lifting speed is not fast enough. In addition, most of them can't save energy. To this end, a novel rotary-lift based hand pallet truck is proposed. Auxiliary and cushioning forces are provided by gas springs. Its lifting height is adjustable. Critically, its lifting mechanism can be locked automatically after ascending or descending into the target positions. Thus, it's easy for just one person to operate. Aiming at saving time, force and energy, a multi-objective optimization design model for the lifting mechanism is built, based on which most key parameters and dynamics indexes can be calculated. Numerical analysis and experiment results show that the proposed model is effective and the design scheme of the novel hand pallet truck is feasible.

1. Introduction

Hand pallet truck, also named as hand forklift, can move goods to required locations quickly. By virtue of saving labor and working efficiently, it is widely used in logistics, gardening and other fields. It's also important for the ground service support in aviation or aerospace enterprises. Two types of hand pallet trucks are used most commonly. One is based on rotary-lift, such as baggage truck. It generally adopts two wheels, and the lifting height is not fixed. Although it has simple structure and response sensitively, it's inconvenient to control and difficult to carry heavier items, because human force should be applied in both horizontal and vertical directions when transporting goods. Another kind of hand pallet truck uses single-stage plunger hydraulic cylinder to lift cargoes. The hydraulic cylinder needs to be driven by shaking or pedaling, which leads to a relatively slow lifting speed. Besides, it is difficult to be used outdoor because of the small tires.

Literature [1] designed a new lifting and guiding mechanism for forklifts, which not only expands the vision field of the driver but also reduces the whole weight of the vehicle. Literature [2] designed an electric lifting mechanism to facilitate the transfer of wheelchair riders. In the literature [3], a new design method was proposed to obtain the optimum design parameters of the scissor-fork lifting mechanism. Literature [4] presented an optimal design method for lifting mechanism of construction tower crane. Although they are enlightening, they can not be used in hand pallet trucks directly. In addition, the traditional design methods usually ignore the importance to store and utilize nature energy[5], such as the work done by gravity when cargoes falling. It does not save manpower work. In general, too little related improvement work has been done. Thus, it is necessary to design a new hand pallet truck with good maneuverability, high efficiency, energy-saving and green environmental protection [6], especially for the transport of specific items.

2. Proposition of a Novel Hand Pallet Truck

The proposed novel hand pallet truck is composed of three modules as shown in Figure 1.

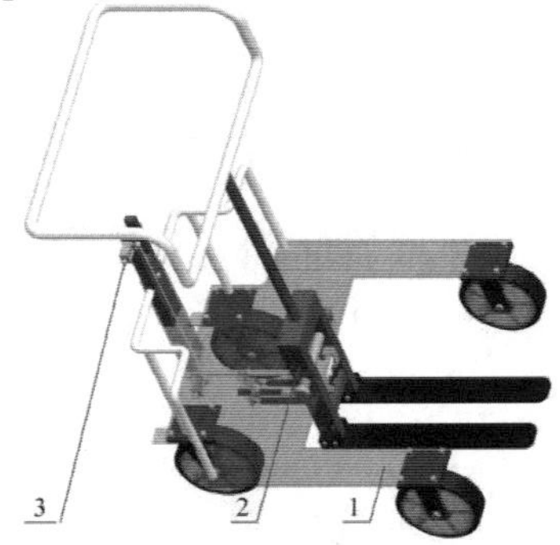

Figure 1. Proposed novel hand pallet truck. 1-truck body; 2-lifting mechanism; 3-joint locking mechanism.

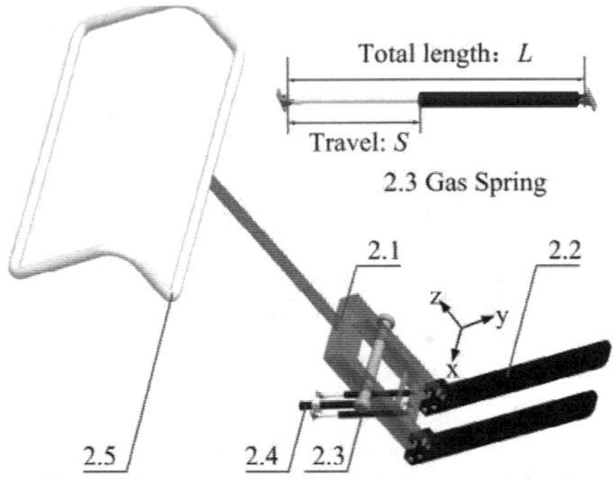

Figure 2. Lifting mechanism. 2.1-rotating frame, 2.2-fork, 2.3-gas spring, 2.4-speed stabilizer, 2.5-handle.

As a transportation platform, truck body mainly includes chassis, lifting height adjusting device, guardrails and wheels. Mounting bases for both lifting mechanism and joint locking mechanism are set under the chassis. The lifting height adjusting device consists of sleeve base, lifting rod and fixing bolt. The sleeve seat is hollow, and there is a threaded hole in the middle of its upper part used to locate the lifting rod. The U-shaped part of the lifting rod can limit the lifting mechanism's position. There are several holes in the middle part of the lifting rod's lower part along the vertical direction. Through adjusting the lifting rod to the desired position, making its appropriate hole aligned with the hole on the sleeve base, then tightening with the fixed bolt, the lifting height can be controlled.

Lifting mechanism is the key function module. Figure 2 shows its composition. Rotating frame is the skeleton of lifting mechanism, which is connected with truck body by rotating axle pin. Its bottom end is connected with two forks by bolts, and they are perpendicular to each other. When not in use, two forks can be folded up to save space. Several rows of holes are arranged below the side of the rotating frame to adjust the initial height of the forks. Down in the middle of the rotating frame, the descending positioning link as well as the fixed plate of the gas springs and the speed stabilizer are welded in turn. By virtue of large elasticity, small size and low price, gas springs are used to provide auxiliary and cushioning force [7]. It consists of cylinder, piston rod and rotary base. Two gas springs are articulated with the truck body and the rotating frame through the rotary bases at both ends. The piston rod should be installed below. As the rotating frame rotates, the piston rotates upward and the cylinder moves axially.The compressed gas springs provide auxiliary force to make the rising process easier. When descending, it acts as a buffer because its acting moment is opposite to that of gravity. Hydraulic speed stabilizer can stabilize the speed of the descending. It produces little stretching force, and the speed can be adjusted. Of course, it could be omitted in common cases.

Joint locking mechanism is used to fix the lifting mechanism. Figure 3 shows its composition. All the three lock switches of it are similar in structure. They are basically made up of positioning pin, outer frame and spring. The two ends of the connection plate are connected with the positioning pin of the down lock switch. Initially, springs are all extended, the forks are horizontal and fixed by the down lock switch. (a) When the pedal is pressed downward, the positioning pin of the lock switch limits the pedal link to the lowest position, the connection plate is raised, the positioning pin of the down lock switch is moved upward, thus the rotating frame is released. (b) When pulling the handle, the rotating frame starts to rotate and then contacts the positioning pin of the up lock switch, which makes the positioning pin to be extruded and move away. It is then jammed after fully passing through the positioning pin, as the positioning pin restores driven by the spring at the same time. Meanwhile, the pedal link is released because its lock switch is driven by the pulley assembly connected with the up lock switch. (c) Through pulling the handle at right end of the up lock switch, the rotating frame can be released and descend to the vertical position, and then will be jammed by the positioning pin of the down lock switch, which keeps the fork horizontal.

3. Multi-objective Optimization Design Model for the Lifting Mechanism

To manufacture the novel pallet truck, it's very important to determine the size, installation position and mechanical parameters of each component, especially for designing the lifting mechanism.

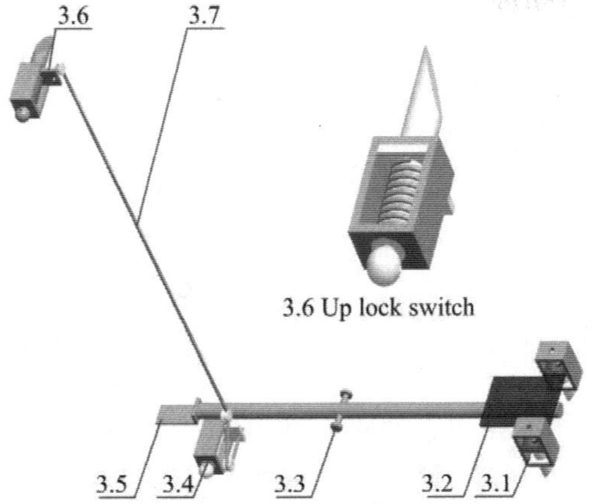

Figure 3. Joint locking mechanism. 3.1-down lock switch, 3.2-connection plate, 3.3-rotating shaft, 3.4-pedal link lock switch, 3.5-pedal link, 3.6-up lock switch, 3.7-pulley assembly.

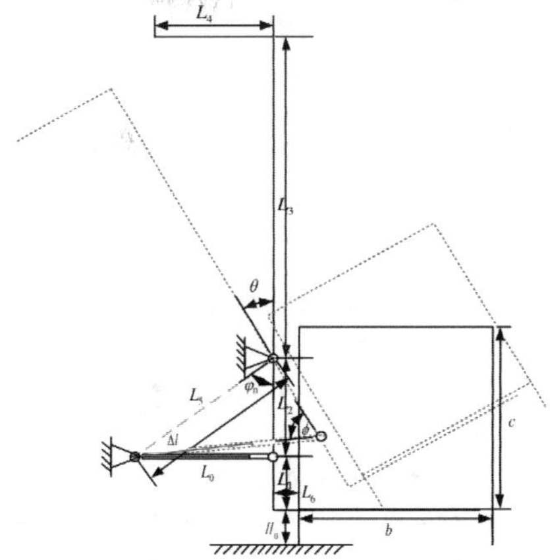

Figure 4. Dimension of the lifting mechanism.

3.1 Determining the size of rotary frame

As shown in Figure 4, L_1, L_2, L_3, L_4, L_6 are respectively denoted as the length from the bottom of the rotating frame to the rotary base of the cylinder, the length from the rotary base of the cylinder to the shaft of the rotating frame, the length from the shaft to the top of the handle, the length of the handle's profile, the horizontal length between the cargo and the rotating frame. Φ, H_0, H_{m1}, H_{m2} and ΔH are respectively denoted as the angle from the vertical position to the target position for the rotating frame, the height between the upper surface of the forks and the ground when the forks are horizontal, the maximum height of the handle in the design of the truck, the ideal height from the handle to the ground, the allowable deviation of H_{m2}. Assuming that the cargo is a homogeneous cuboid, the length, width and height are denoted as *a, b* and *c* respectively. Limited by wheels size and other constraints, L_1, L_2, L_6, H_0, H_{m1}, H_{m2} and ΔH are deterministic. Longer L_3 and smaller Φ can lead to smaller human force and less work. Hence, other parameters can be got at the boundary of the constraint conditions.

3.2 Determining the specification of gas spring

Let L_0 denote the length of the gas spring when the fork is horizontal, L_S denote the length of the gas spring extending out after the cargo is lifted to target position, L_5 denote the distance between the left rotary base of the gas spring and the shaft of the rotating frame, φ_0 denote the angle between L_5 and the vertical direction. According to the cosine theorem, the following relationship is established:

$$L_0=\left(2L_S-2L_2\sin\Phi\right)^{-1}\left[2L_2^2\left(1-\cos\Phi_2\right)-L_S^2\right] \quad (1)$$

The size of gas spring mainly includes total length *L* and stroke *S*, and usually need to be selected from the product list. Here, they are defined as a set of gas spring sizes *A* containing different numerical pairs *(L, S)*. In addition, to ensure safety, the gas spring should have enough safety margin during compression or stretching, which is defined as Δ_S. The optimal value of L_0, L and S can be obtained by solving the following optimization problems by using classical linear search algorithm.

$$\begin{aligned} &P_1:\min_{L_S,\,L,\,S}\quad L_0,\ L,\ S \\ &s.t.\ C_1:L-S+\Delta_S\le L_0,\ C_2:S-2\Delta_S\ge L_S,\ C_3:\left(L,S\right)\in A \end{aligned} \quad (2)$$

Next, to determine the elasticity of the gas spring, the optimization method is used. In reality, just 1-2s is needed to lift the cargo. We think that the rising time expressed as T increases linearly with the cargo weight, and the aim to save lifting time can be satisfied as soon as it does not exceed a maximum value. The second goal of labor-saving can also be gotten if only the maximum human force required does not exceed a certain bearing range. The ultimate goal is to minimize human effort. In general, the above three physical quantities are negatively correlated with the elasticity of the gas spring, so the greater elastic force, the better performance of the lifting mechanism. However, the elasticity cannot be arbitrarily large, and certain constraints should be met. We denote f_h, F_S and G respectively as human force, gas spring elasticity and gravity. $f_{h\max}$, $F_{hl\max}$, $F_{hp\max}$, W_{f_h}, $T_{\max}$, Φ_1 are respectively denoted as the maximum human pull force during the rising procedure, the maximum human force human can afford to lift the handle upward without payloads, the maximum pull force human can bear, the work done by human force during the rising procedure, the maximum lifting time, the angle between gas spring and the rotating frame after the cargo is lifted in place. We denote r, θ, γ_0, ϕ, θ_1, Δl respectively as the distance between the cargo center of mass and the rotating axis of the truck, the rotation angle of the rotating frame, the angle between $\bar{r}$ and the vertical direction when θ is equal to 0, the angle between the gas spring and the rotating frame, the rotary angle of the gas spring, the expansion length of the gas spring. Thus, the multi-objective optimization model is built as follows:

$$\begin{aligned} P_2 : &\min_{T,F_s} \quad f_{h\max}, W_{f_h} \\ s.t.\ & C_1 : F_s L_2 \cos(\Phi - \Phi_1) \le G\left(\left(2^{-1}b + L_6\right)^2 + \left(L_1 + L_2 - 2^{-1}c\right)^2\right)^{1/2} \sin(\gamma_0 + \Phi), \\ & C_2 : F_s L_2 \le F_{hl\max} L_3,\ C_3 : f_h \le F_{hp\max}, \\ & C_4 : T \le T_{\max},\ C_5 : \theta \le \Phi,\ C_6 : F_s, \theta > 0 \end{aligned} \tag{3}$$

This optimization problem can also be solved by using search algorithm.To simplify the calculation, we assume that the cargo rotates at uniform acceleration during the first 0.5Φ with acceleration α_0 and then at uniform speed at last 0.5Φ with angular velocity ω_1. The following equation is gotten:

$$\alpha_0 = \left(4T^2\right)^{-1} 9\Phi,\quad \omega_1 = 3\Phi(2T)^{-1} \tag{4}$$

The mean angle of the direction of human force deviating from the vertical direction of the rotating frame is set as δ. f_h can be calculated as below:

$$f_h = \begin{cases} (Gr - F_s L_2 \sin\phi + I\alpha_0)(L_3 \cos\delta)^{-1}, & \theta \le 2^{-1}\Phi \\ (Gr - F_s L_2 \sin\phi)(L_3 \cos\delta)^{-1}, & \theta > 2^{-1}\Phi \end{cases} \tag{5}$$

$$r = \left(\left(2^{-1}b + L_6\right)^2 + \left(L_1 + L_2 - 2^{-1}c\right)^2\right)^{1/2} \sin(\gamma_0 + \theta) \tag{6}$$

$$\gamma_0 = \operatorname{arccot}\left(\left(L_1 + L_2 - 2^{-1}c\right)\left(2^{-1}b + L_6\right)^{-1}\right) \tag{7}$$

$$\theta_1 + 2^{-1}\pi = \theta + \phi \tag{8}$$

$$\phi = \arccos\left(\left(L_2^2 + (L_0 + \Delta l)^2 - L_5^2\right)\left(2L_2(L_0 + \Delta l)\right)^{-1}\right) \tag{9}$$

$$\Delta l = \left(L_2^2 + L_5^2 - 2L_2L_5 \cos\left(\arctan\left(L_0 L_2^{-1}\right) + \theta\right)\right)^{1/2} - L_0 \tag{10}$$

Let W_{F_S} denote as the work done by gas springs. Δh and I are respectively denoted as the lifting height and the moment of inertia of the cargo. Because it is difficult to be calculated with an analytic expression, W_{F_S} can be obtained by adapting numerical calculation method. According to the kinetic energy theorem, the work done by manpower is expressed as follows:

$$W_{f_h} = G\Delta h - W_{F_S} + 2^{-1} I \omega_1^2 \tag{11}$$

$$\Delta h = \left(\left(2^{-1} b + L_6 \right)^2 + (L_1 + L_2 - 2^{-1} c)^2 \right)^{1/2} \left| \cos\gamma_0 - \cos(\gamma_0 + \Phi) \right| \tag{12}$$

$$I = \frac{G}{3gc} \left(\left(b^2 + 3bL_6 + 3L_6^2 \right) c + \left(c - L_1 - L_2 \right)^3 + \left(L_1 + L_2 \right)^3 \right) \tag{13}$$

$$W_{F_s} = \sum_{i=1}^{N} F_s L_2 \cos\left(i\Phi N^{-1} - \theta_{1i} \right) \Phi N^{-1} \tag{14}$$

4. Numerical Simulations and Mechanics Experiments

Table 1 and Table 2 respectively show the initial parameters of the lifting mechanism for the case of 1600N cargo load, and the optional specifications of gas spring. Sizes of the rotating frame and the handle are calculated firstly. ***L_3* is 1300mm, *L_4* is 420mm**. We consider that the truck has two lifting heights expressed as $H_{\Delta l}$ and $H_{\Delta h}$. They refer to the distance from the lowest point of cargo feet to the ground. Based on our model, the size of gas spring is gotten according to $H_{\Delta h}$. ***L* is 450mm, *S* is 150mm**. The elasticity of gas spring and relevant kinetic indexes are calculated according to $H_{\Delta l}$. ***The total elasticity of gas springs needed is 1000N***. Using the parameters above, the kinetic indexes with or without gas springs are calculated when the cargo weight changes from 900N to 1600N, as shown in figure 5. The cargo is a homogeneous cuboid with a length of 600mm and a width of 600 mm. Both the centroid height of the cargo and T_{max} vary linearly with the cargo weight.

Table 1. Input parameters

L_1	L_2	L_6	H_0	H_{m1}	H_{m2}	ΔH	$H_{\Delta l}$	$H_{\Delta h}$	Δ_S	$T_{\max}$	$F_{hl\max}$	$F_{hp\max}$	G	g	δ	N
112 mm	208 mm	130 mm	80 mm	1700 mm	1100 mm	210 mm	80 mm	185 mm	10 mm	1.6 s	160 N	320 N	1600 N	10 m/s^2	0.26 rad	100

Table 2. Optional specifications of gas springs

L (mm)	340	400	537	580	200	250	300	350	390	**400**	450	470	480	500	550	600	650	700	765
S (mm)	130	150	225	245	60	85	110	130	150	**150**	180	190	190	205	230	255	280	305	300
Elasticity (N)	100-1400				100-4500														

A prototype was made. Its mechanical properties were tested by SH-III-500N push-pull force tester. The human pull force measured can be output to the software in the host computer in real time through USB for display and storage, as shown in Figure 6 (a). With the cargo weight changing, the human force required for lifting was measured under the conditions with or without gas springs, as shown in Figure 5 (a, b). The human force required for horizontal walking was also measured on a site about 20m long. The statistical result is shown in Figure 6 (b). Figure 5 (a) shows the human force in the process of rotating ascending varying with time when *G* equals 1600N. Figure 5 (b) and (c) show the average human force and the work by manpower varying with cargo weight. They show that about 150N and 105J can be saved with the help of gas springs. Without gas springs, the work done by manpower is more than that of gravity, because there's still kinetic energy when the cargo getting to the target height. The analytical results of the proposed model are in good agreement with the experimental results. The truck weight without payload is about 1600N. When the cargo weight increases, the required horizontal force increases from 83N to 115N, which can be tolerated.

5. Conclusion and Future Work

To address the problems, such as slow lifting speed and unstable operation, a novel rotary-lift based hand pallet truck assisted by gas springs is designed, a multi-objective optimization model for the design of lifting mechanism is proposed, based on which a functional prototype was made. The results of numerical calculation and mechanics experiment show that the proposed model is basically consistent with the actual situation. The design goal of saving lifting time, human force and work can be achieved. The novel truck is convenient, safe, reliable and green. However, it can not be used to carry goods requiring limited sloshing, and more useless work is done due to the inherent defects of rotary lifting mode. Next, we will design a horizontal lifting truck assisted by gas springs.

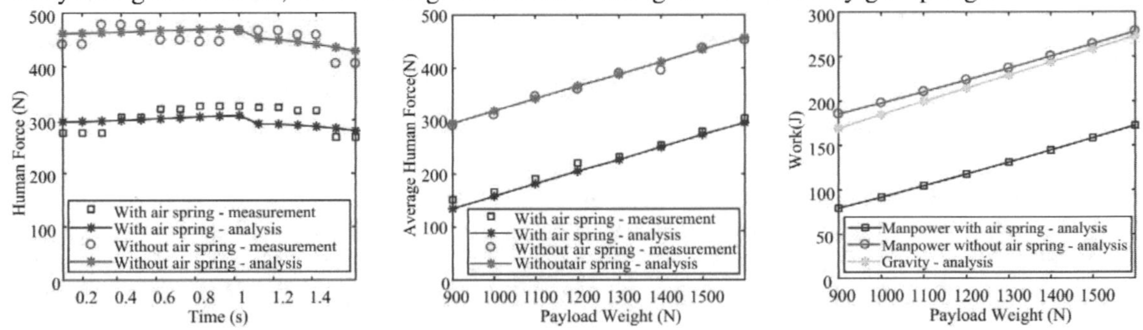

(a) Human Force *(G = 1600N)* (b) Average Human Force (c) Work

Figure 5. Mechanical properties comparison with or without air spring

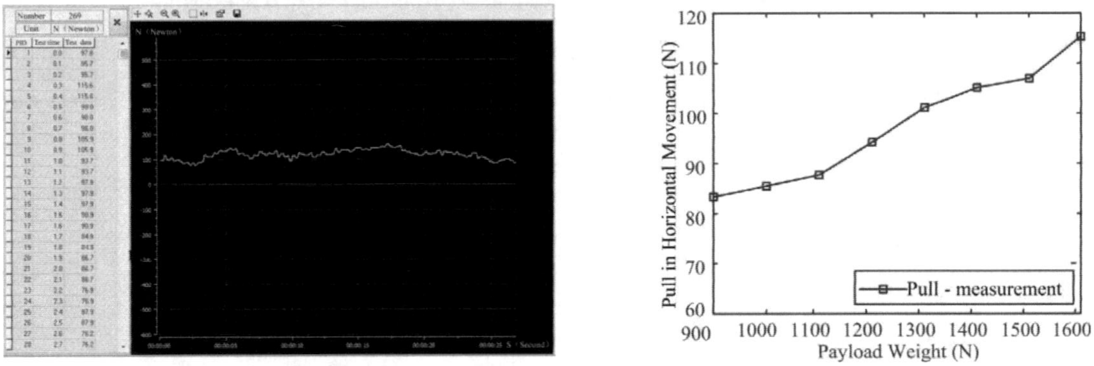

(a) Payload equals 1600N (b) Payload varies from 900N to 1600N

Figure 6. Pull force in horizontal movement

References

[1] Zhang W, Zhang X, Yan C, Xiang S and Wang L 2015 A characteristic triangle method on input vectors of scissor lift mechanism and its applications in modeling and analysis *J. Adv. Mech. Des. Syst.* **9** 1-20.

[2] Wang J, Zhao J, Chu F and Feng Z 2010 Innovative design of the lifting mechanisms for forklift trucks *Mech. Mach Theory.* **45** 1892-1896.

[3] Jiri Stastny and Vit Motycka 2018 Design Optimization of lifting mechanisms *WMCAUS 2018. Prague, Czech Republic*. 1-8.

[4] Yoshikazu Mori, Norikatsu Sakai and Kaoru Katsumura 2012 Developmentof a Wheelchair with a Lifting Function *Adv. Mech. Eng.* **2012** 1-9.

[5] T.A. Minava, A. Virtanen, L. Laurila and J. Pyrhönen 2012 Storage of energy recovered from an industrial forklift Automat. Constr. **22** 506-515.

[6] Smith S and Hung P Y 2015 A novel selective parallel disassembly planning method for green design *J. Eng. Design.* **26** 1-19.

[7] Ou X, Song J, Lu Q, Hu C and Wang Q 2019 Analysis and selection of gas spring of closet-type 0-90° foldable bed Chin *J. Beijing. Forestry. Uni.* **41** 147-153.

IOP Conf. Series: Materials Science and Engineering **816** (2020) 012015 doi:10.1088/1757-899X/816/1/012015

Development Research and Crucial Technology Analysis of Scaled 3-Bearing Swivel Duct Nozzle Rotary Drive System

QI Hao[1], YANG Xiao[2], WANG Zehe[1], ZHU Huajuan[3]

[1]College of Mechanical and Electrical Engineering, Agricultural University of Hebei, Baoding 071001, China.
[2]College of Computer Science and Technology, Tsinghua University, Beijing, 100084, China
[3]College of Economics and Management, Agricultural University of Hebei, Baoding 071001, China

Email: [1]365145900@qq.com; [2]yangxiao502@126.com; [3]793139070@ qq.com

Abstract. The 3-bearing swivel duct (3BSD) nozzle for short-range/vertical take-off and landing(S/VTOL) fighters is a crucial component for aircraft to achieve vertical and/or short take-off and landing, which significantly improves the manoeuvre ability of the aircraft, while the nozzle rotary drive system is the crucial technology for S/VTOL functions which has significant impact on the fighter performance. Based on the principle and characteristics of the 3BSD nozzle, this paper systematically reviews the development and evolution of the 3-bearing swivel duct nozzle, and analyses and summarizes the types and characteristics of the 3BSD nozzle rotary drive system. Three crucial technologies of the 3BSD nozzle rotary drive system are evaluated and discussed in detail. Suggestions on future research and development of equipment are given.

1 Introduction

S/VTOL aircraft combines the advantages of fixed-wing aircraft and rotary-wing aircraft. It can significantly improve the aircraft's the take-off and landing ability to reduce the requirements for the take-off and landing environment, the higher flight speed can improve manoeuvre ability as well. From theory to practice, it has gone through more than 70 years of development. At present, the main types of its power plant are: lift engine, thrust vector nozzle engine, lift fan system, spiral rotor system, and stallable or retractable rotor system [1-6].

The thrust vector direction of the thrust vector nozzle engine can be changed by rotating the nozzle angle. The steering mechanism of this engine exhaust nozzle is represented by the British Harrier fighter, the Yak series of Russia, and the F-35B fighter of the United States. According to the current practice [7,8], in terms of lightweight, high efficiency, stealth performance and control technology, the 3BSD nozzle can meet the requirement of S/VTOL propulsion system.

S/VTOL aircraft using thrust vectoring not only has a conventional aerodynamic rudder surface, but also uses large angle thrust vector and various forms of lift devices to provide direct lift. In order to meet the different thrust direction requirements, the S/VTOL aircraft needs to be deflected by more than 90°. The nozzle end rotation stroke reaches 180°, and the 3BSD nozzle rotation angle has a strong nonlinear relationship with the thrust vector rotation angle. Therefore, the 3BSD nozzle rotary drive system is considered as a crucial technology that directly affects S/VTOL aircraft performance.

The main purpose of this paper is to review the characteristics and development of the 3BSD nozzle rotary drive system basing on its basic principle and characteristics. The layout of this study is organized as follows: Section 2 is aimed at explaining the principle and characteristics of the 3BSD nozzle, and then, the research status of it is presented in Section 3. In Section 4, various of key technologies of each type of 3BSD nozzle presented in the former section are discussed. At the end of the Section 4, these types of BSD nozzle are compared on basis of their beneficial characteristics as well as their drawbacks. In the last section, the future development on 3BSD nozzle is discussed.

2 Principle and Characteristics of Scaled 3-Bearing Swivel Duct Nozzle

2.1 Structure

The 3BSD nozzle is mainly composed of three nozzle ducts and rotary drive system, between nozzle ducts are connected by a thin-walled ring bearing. The end section of the first duct inlet is circular, the inlet of 3BSD nozzle first duct is fixed with the outlet of the engine, the end face of the outlet is circular too, and the diameter is the same as the diameter of the inlet end section of the second duct. The second duct is an elliptical tube. Both the outlet end sections of the second duct are circular. Meanwhile, there is an angle between the elliptic axis and the normal of oblique section. The end section of the third duct is round, and the diameter of the outlet of the second section is the same.

In normal flight mode, the three-ducts keeps the centre line coincident and the thrust is behind the engine axis. In vector flight mode, the rotary drive device drives the nozzle to rotate, changing the nozzle orientation.

2.2 Law of movement

During the movement of the 3BSD nozzle, the three-duct nozzle is sustained in a certain fixed plane. the relationship between the rotation angle of each nozzle and the total vector angle is expressed as follows [1]:

$$\Omega_1 = -arctan(tan(\Omega_2/2)cos\,\theta) + \delta_{Ny};\ \Omega_2 = arccos\left(\frac{cos\,(\delta_N/2) - cos^2\theta}{sin^2\theta}\right);\ \Omega_3 = -\Omega_2.$$

where Ω_1, Ω_2 and Ω_3 denote the three angles input quantities of the 3BSD nozzle, δ_N and δ_{N_y} denote the longitudinal and lateral deflection angles of the nozzle, θ denotes the angle formed by the nozzle end section and the pipe cross section. When $\Omega_2=\pi=-\Omega_3$ the maximum vector angle is reached: $\delta_N=4\theta$ [1].

3 Overview of the Development of 3BSD Nozzle

3.1 Foreign typical 3BSD nozzle

In 1960, Jack [9] first proposed the concept of a rotating nozzle that mounted a gear on the outer wall of the nozzle and rotated it through the gear drive.

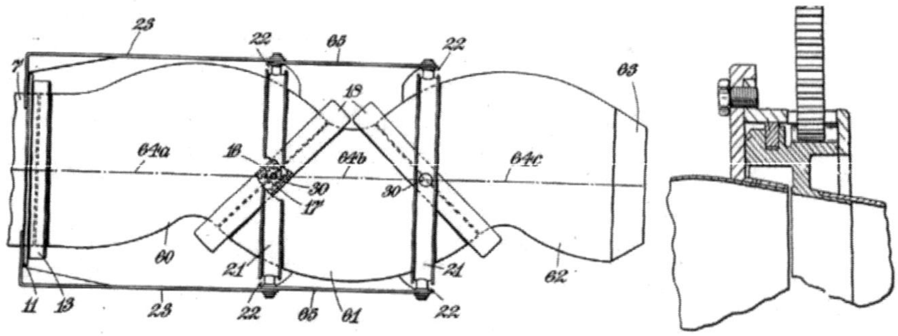

Figure. 1. Jack three-stage rotary nozzle structure

In 1966, the motor used in the multi-section wedge nozzle developed by Carroll [10] has two output shafts, which drive the gears to drive the synchronous rotation of the two bearings of the second and third duct.

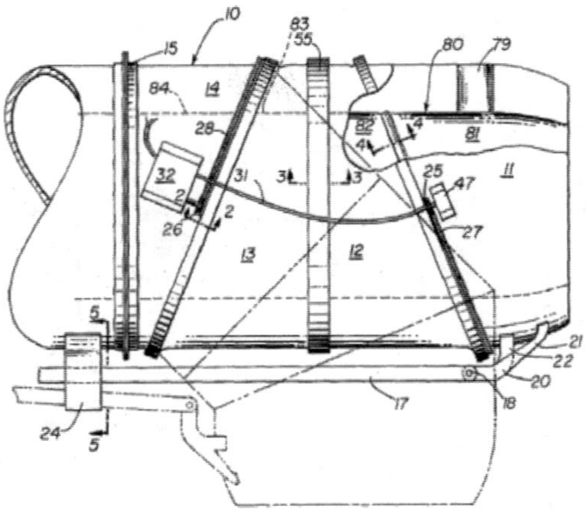

Figure. 2. Carroll multi-stage rotary nozzle structure

In 1969, the nozzle proposed by Gerhard [11] was composed of four straight sections with irregular curved surfaces. The nozzles of each section were connected by bearings. Each section of the nozzle was equipped with gears to ensure the synchronous deflection of each upper and lower section.

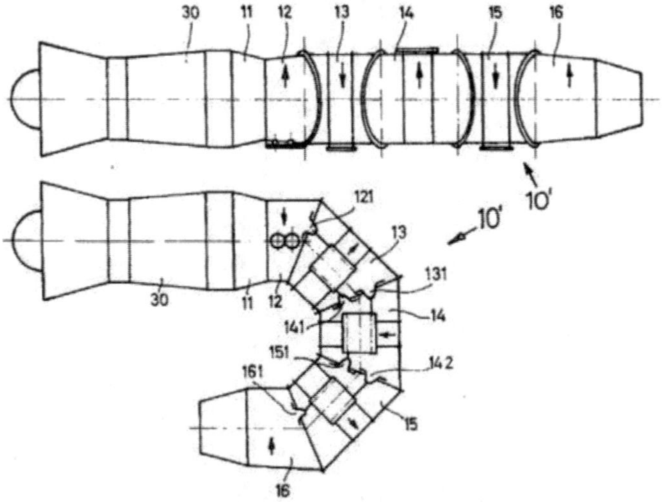

Figure. 3. Gerhard four-stage rotary nozzle structure

Later this year, Alexander [12] extended the maximum rotation angle of the nozzle to 90°, and added a reverse device to ensure the same speed rotation through the gear shaft.

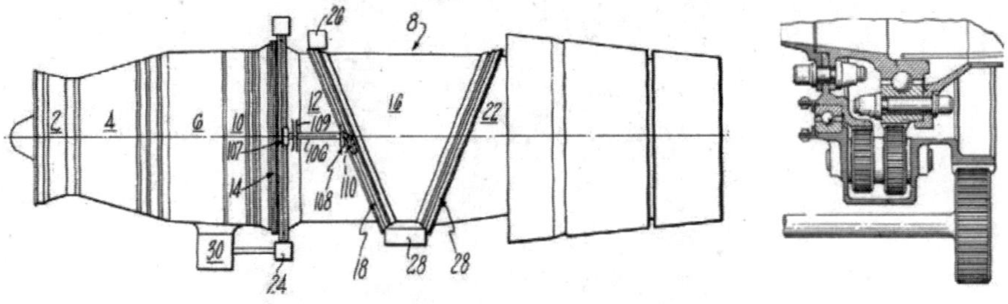

Figure. 4. Alexander 3BSD nozzle gear drive system

In 1972, Dudley [13] replaced the gear shaft and bevel gear assembly with a more compact ring structure.

IOP Conf. Series: Materials Science and Engineering **816** (2020) 012015 doi:10.1088/1757-899X/816/1/012015

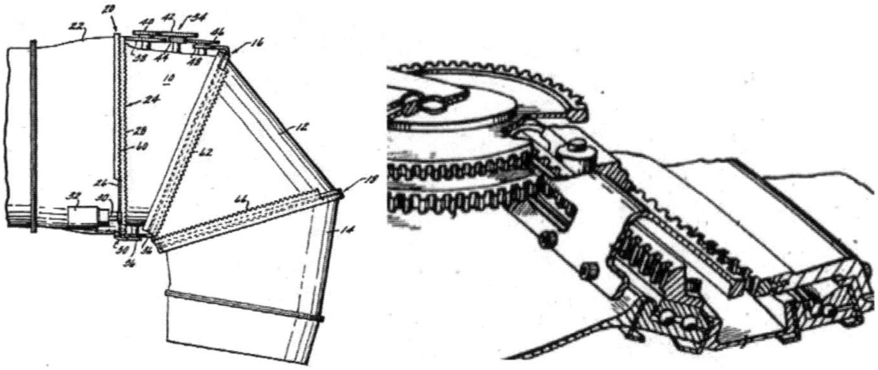

Figure. 5. Dudley 3BSD nozzle multi-gear drive system

In the scheme designed by Michael C. Roberts et al [14], two fixed pistons on the outer wall of the inner tube and the fixed piston cylinder on the inner wall of the outer tube were divided into three circular arc independent spaces. The rotation of the inner tube relative to the outer tube is achieved by the valve body controlling the high-pressure fluid to push the floating piston.

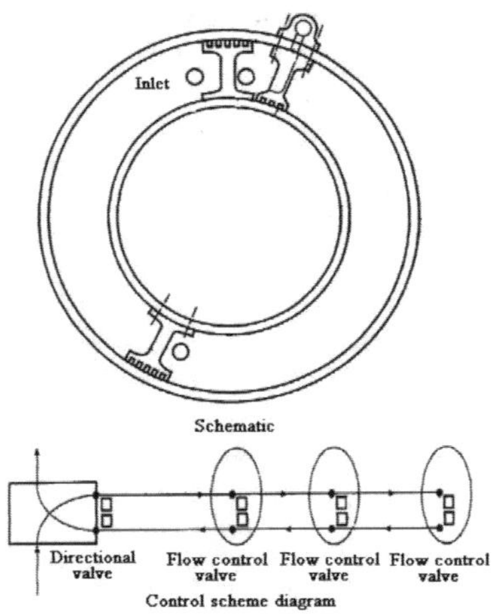

Figure. 6. Michael liquid / pneumatic thrust vector nozzle

The first practical application of the 3BSD nozzle was the main engine R-79V-300 of the former Soviet "Yak-141" SVTOL aircraft. When the short-range aircraft takes off, the nozzle rotates downward by 63°; when it takes off vertically, the nozzle rotates downward by 95° [15].

Figure. 7. "Yak-141" R-79V engine

The first duct of the F-35B is controlled by the flight control system to control the gear deflection of the rotating part of the aero-motor drive, while the second and third duct of the nozzle are driven by the hydraulic system, the 3BSD nozzle can be rotated from 0° to 95° in 2.5s [16].

Figure. 8. F135-PW-600 engine assembly used in F-35B

China's attention to S/VTOL aircraft started long time ago. In 1969, 601 institute included S/VTOL flapping variable-wing aircraft into special research projects. China's new round of S/VTOL aircraft pre-research has been carried out for at least a decade, but when compared with foreign research, domestic research on 3BSD nozzles started late, and the research level is low which mainly focus on the verification of the principle of S/VTOL aircraft aimed at F-35B in scientific research institutes and universities.

The 3BSD nozzle developed by Northwestern Polytechnical University [3] drives three ducts through three sets of motors to achieve vector rotation around three axes.

Figure. 9. 3BSD nozzle of Northwestern Polytechnical University [3]

Tsinghua University [17] has manufactured the largest S/VTOL demonstration named THU-F-35B in China. The nozzle rotary drive system adopts the first domestic "8" font-type precision cable drive, transmitting torque in the form of friction to drive the nozzle to the desired angle and change the direction of deflection. Through optimisation of the structure, the number of minimum drive motors required to deflect the 3BSD nozzle thrust vector is reduced by one compared with conventional single-stage drives.

Figure. 10. 3BSD nozzle model of Tsinghua University

Lanzhou Jiaotong University [18] also adopts the cable transmission mechanism. The three drive motors are axially connected with reducer and fixed to the outer ring of each duct bearing through the end flange. As the driven wheel ring is uniformly opened with a wire rope guiding groove.

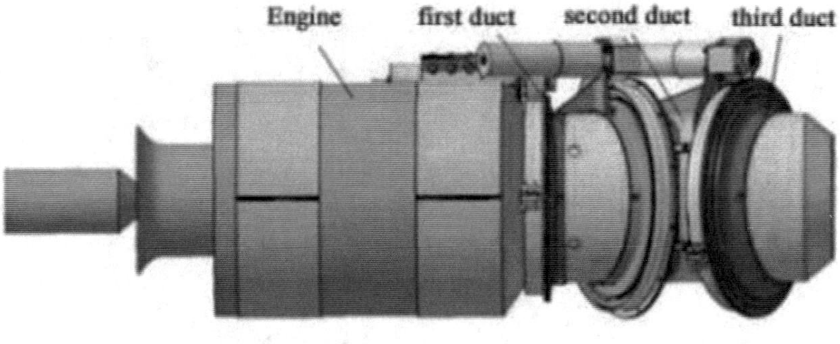

Figure. 11. 3BSD nozzle model of Lanzhou Jiaotong University

AVIC Engine has carried out the principle verification of the S/VTOL aircraft propulsion system. The 3BSD nozzle rotary drive scheme is shown in Figure 12:

Figure. 12. 3BSD nozzle model of AVIC Engine Institute

4 Crucial Technical Analysis

4.1 3BSD nozzle rotary drive system working requirements

The saturation speed of the actuator is an important parameter of the aircraft. The 3BSD nozzle has a much longer travel range than conventional actuators. The rotation range for a given nozzle deflection command(δ_N , δ_{Ny}), Ω_1 , Ω_2 and Ω_3 is different [16]. Therefore, some actuators may saturate when other actuators dose not.

Firstly, the relationship between rate saturation speed of Ω_2 and δ_N is analysed. Secondly, Ω_3 is taken into account. Finally, Ω_1 and deflect directions are included. The second equation in the above equations can be rewritten as:

$$\cos^2(\theta)+\sin^2(\theta)\cos\Omega_2=\cos(\delta_N/2)$$

By differentiating both sides, the relationship between the speed of Ω_2 and δ_N can be described as:

$$\dot{\delta}_N=\dot{\Omega}_2/d\Omega_2/d\delta_N$$

The function of $d\Omega_2/d\delta_N$ is shown in Fig. 13.

In order to satisfy the relationship between the rotation angle of each ducts, the thrust vector force and the maximum vector angle, the speed constraint imposed on the second rotation pair is:

$$max\left\{-1.8\dot{\Omega}_{1\min},\dot{\Omega}_{2\min},-\dot{\Omega}_{3\max}\right\}\le\dot{\Omega}_2\le min\left\{-1.8\dot{\Omega}_{1\max},\dot{\Omega}_{2\max},-\dot{\Omega}_{3\min}\right\}$$

Suppose δ_{Ny} is a constant, the equation could be deduced:

$$\dot{\Omega}_2 = \dot{\Omega}_1 d\Omega_2 / d\Omega_1$$

The function is shown in Fig. 14.

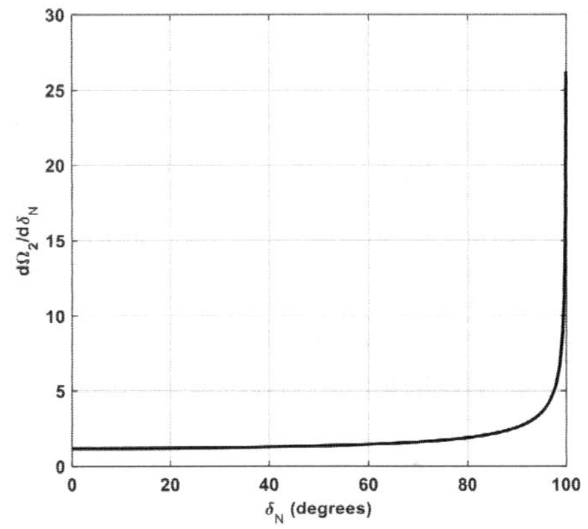

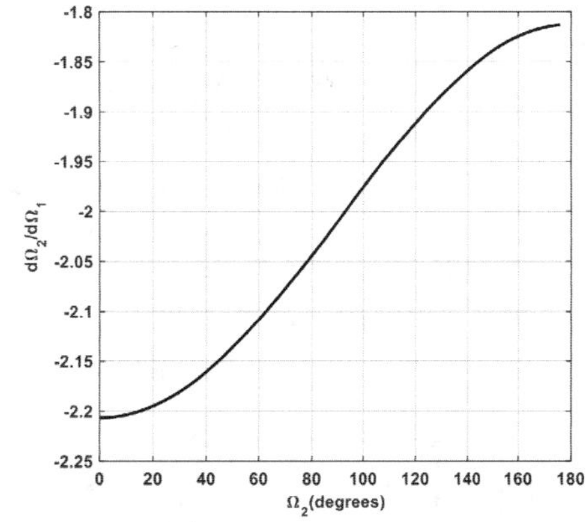

Figure. 13. $d\Omega_2 / d\delta_N$ curve **Figurer. 14.** $d\Omega_2 / d\Omega_1$ curve

Since the deflection speed of the 3BSD nozzle at different positions is limited, we can deduce that a relatively small rotation angle of Ω_1 can achieve a given lateral control force T_{Nv} . In normal flight mode, T_{Nz} close to 0 at most times, at this time, the corresponding δ_{Nv} will become infinity. Considering that the rotational angle is limited, the rotation angle of Ω_1 is 90°, which makes the flight control system unusable for lateral manipulation of the normal flight mode. Based on the above analysis, the 3BSD nozzle can provide lateral force/torque manipulation in hover and transition modes, while lateral force/torque manipulation is not available in conventional mode.
According to the existing research theory and results, it can be concluded that the 3BSD nozzle rotary drive system must meet the following conditions:
1.During the rotation of the nozzle, the thrust vector is maintained in a vertical plane;
2.The vector change process thrust is continuous;
3.The second rotating pair and the third rotating pair rotate in opposite directions at the same speed;
4.Complete the maximum vector angle change within 2.5s;
5.The axial distance of the nozzle is as short as possible;
6.The maximum vector change angle of the nozzle is ⩾95°, and should not be greater than 100°.

4.2 Driving Mode

Table 1.3BSD nozzle rotary drive system form, characteristics and drawbacks.

Solutions	Research Unit	Characteristics	Drawbacks
gear	Jack	Pipe outer fixed gear	Pneumatic load acts on gear
	Carrol	the bearing is selected as the connecting part for the first time, and the synchronous rotating device is provided.	The mounting position of the motor causes one of the two output shafts of the motor to be bent
	Gerhard	The outer side is equipped with a gear for synchronously deflecting the upper and lower stages	axial dimensions increased resulted in aerodynamic losses
	Alexander	Change the maximum vector angle from 180° to 90°	the gear shaft is used as the linkage mechanism, unable to guarantee the axes of the segments in the same plane when the nozzle rotates,
	Dudley	Replace the gear shaft and bevel gear device with a ring gear structure	The use of planetary gear trains to increase the weight is not conducive to the lateral force of the nozzle
	NPU	Single-stage control, external ring gear drive	No synchronizing device, single-stage drive, integrated ring gear is not easy to repair
	AVIC Engine	Adjustable motor position	No synchronizing device, single-stage drive,

	Institute	according to nozzle vector angle	heavy weight of the drive system
Liquid/pneumatic	Michael C. Roberts	Cancel the sync device, the hydraulic/pneumatic drive is adopted to rotate nozzle	Leakage and inflammability of The hydraulic system; low transmission precision of pneumatic system due to the compressibility of air
	Yak-141	Single-stage control, external ring gear drive	Each segment is driven separately and the drive system is heavy
	F-35B	Combined with aviation steering gear and segmented hydraulic drive nozzle	The hydraulic system is heavy and the pipeline is complicated, and the intermediate pipeline needs to be bent.
Cable	LJU	Rotating the nozzle in the form of friction by the wire rope	No synchronizing device, single-stage drive, heavy weight of the drive system
	THU	the "8" font-type winding method, only one motor is used to control the deflection of the second and third of rotating pairs of the nozzle	This transmission method has only been tested on verification machine reduced in proportion and not sure whether it is suitable for real models with large weight.

The 3BSD nozzle rotary drive system not only needs to meet the requirements of the change of thrust vector angle of 3BSD nozzle, but also consider the overall aerodynamic performance, lightweight, high precision and heat resistance requirements of the aircraft. Comprehensively comparing the three development directions of 3BSD nozzle rotary drive system, we can find that although the hydraulic drive scheme has been applied to the F-35B and other models, there is still room for improvement in system complexity, heavy weight and fuel safety; as one of the most widely used transmission methods, gear transmission is the most mature technology. However, the problems of gear dynamic load, elimination of transmission clearance and integrated control connection of two and three stage nozzles still need to be solved. Flexible cable transmission scheme, as a new type of high precision and lightweight transmission scheme, is simple, efficient, accurate and reliable, without dynamic load. Although the actual application effect needs to be further verified, it is a good direction choice for the present domestic research stage.

4.3 Fly/push integrated control

The S/VTOL aircraft dynamics model is a constrained nonlinear system, and the actuators are redundant and heterogeneous. The transition process control requires the aircraft terminal state to meet the constraints, and the integrated control of the aircraft and power plant is complicated. Advanced S/VTOL aircraft power systems and jet airflow effects are more complex than traditional fighters, and the steering modes are more diverse, making linear controllers difficult to apply.

For the requirements of aircraft control and engine control, as shown in Figure 15, Pratt & Whitney [19] developed a multivariate control system based on the principle of partial dynamic transformation and model predictive control for the F135-PW-600 engine. This control system decomposes the complex, cross-linked and non-linear propulsion system dynamics model and presents the engine as a flight control virtual actuator. The entry is a module that decomposes and analyzes flight control commands. The multivariate limit module is used to determine that the target and limit are valid in the case of a given correction, and complete the corresponding dynamic transformation, nonlinear engine model and linearization. The engine is used to predict the variables that cannot be measured and to perform dynamic transformation. The estimator correction module is used to compare the feedback measurement and the model prediction results to indicate the mode and output deviation of the drive axis and compensate the calculation error of the model.

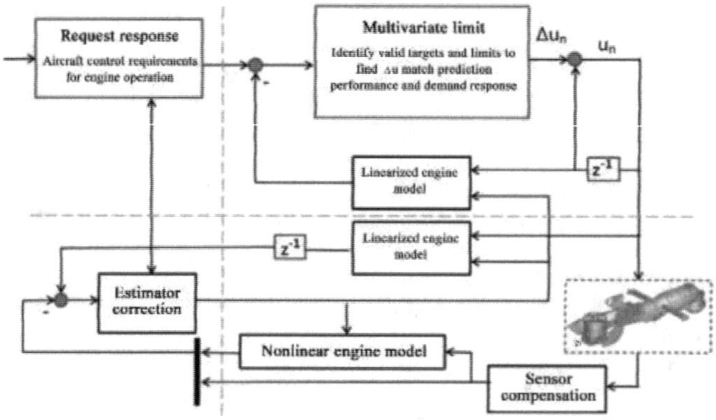

Figure. 15. Multi-variable control system of F135-PW-600 engine developed by Pratt & Whitney

The relationship between the three-duct nozzle angle and the vector thrust deflection angle is a monotonic strong nonlinear relationship, and the rotation sub-stroke is long during the use of the nozzle. Therefore, the study of the nozzle dynamic characteristics requires consideration of the three-stage actuator rate constraint and the rotation pair dynamic and rotational synergy.

Tsinghua University designs the nozzle coordination controller according to the inverse kinematics control law. The structure of the controller is shown in Fig.16. The dynamic allocation method is used to generate the tracking commands of the three actuators, Ω_{1c} , Ω_{2c} and Ω_{3c} the dynamic allocation method generates the three-level rotation sub-commands by nozzle inputting command δ_N and δ_{Ny} . The nonlinear kinematics relationship of the nozzle will bring difficulties to the design of the dynamic allocation method. This project replaces the control target and achieves the purpose of the variable nonlinear distribution assignment problem as a linear control assignment problem. Through the nozzle kinematics model, Ω_1 and Ω_2 are obtained, and the kinematic relationship" $\Omega_2 = -\Omega_3$ "is selected as the constraint equation. The optimization objective function for control allocation is:

$$J = \|\Omega_2(k+1) - \Omega_{2des}(k+1)\|_2 + \gamma\|\Omega_1(k+1) - \Omega_{1des}(k+1)\|_2$$

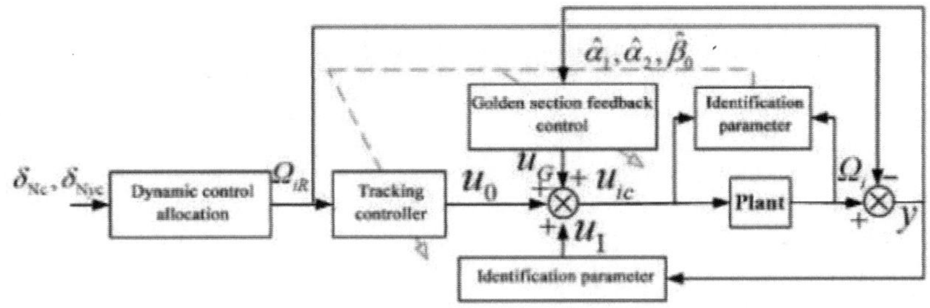

Figure. 16. Coordinated control method structure diagram

By solving the optimization problem, a three-stage rotation sub tracking instruction, Ω_{1R} , Ω_{2R} and Ω_{3R} can be obtained.

The principle of the nozzle coordination control method to achieve the system is shown in Figure 17. Among them, δ_{Nc} 、 δ_{Nyc} is the nozzle command signal, T_{Nc} is the engine thrust command signal, the DSP controller calculates the nozzle command signal and sends it to the nozzle servo through the command signal and the three-stage rotation of the nozzle. The location is sent to the control computer via the DSP controller and stored.

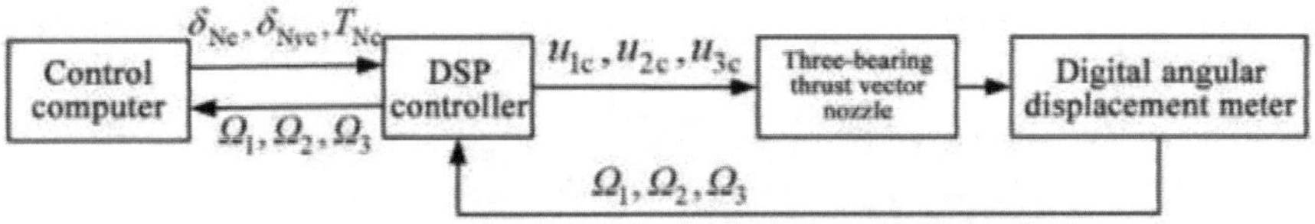

Figure. 17. 3BSD nozzle coordination controller implementation block diagram

4.4 Characteristic test

In addition to obtaining the data of lift, control response, flow field characteristics and other data of the aircraft through the S/VTOL aircraft before the test flight, the static and dynamic response tests such as deflection angle, deflection rate, driving torque and aerodynamics of 3BSD nozzle are carried out mainly through ground ignition test.

Figure. 18. Yak-141 hovering and VAK-191B ground mooring test

IOP Conf. Series: Materials Science and Engineering **816** (2020) 012015 doi:10.1088/1757-899X/816/1/012015

In order to test the dynamic characteristics of the nozzle and establish the dynamic model, the aeronautical vector nozzle test platform basically uses the high-precision six-component balance to record the vector thrust response. Firstly, the static thrust response of the turbojet engine is measured, and the engine dynamics model is established. The dynamic model of the engine is then identified by a dynamic force test. By measuring the static and dynamic response of the 3BSD nozzle separately, the nozzle static thrust deflection model and dynamic model are established in turn. Figure 20 is a schematic diagram of the test model.

The balance is divided into inner balance and external balance in the way of use [20]. When the vector nozzle is tested, there is a high-speed airflow inside, and there is no space for installing the inner balance, so only the external balance can be used. The external balance is mainly divided into two categories: mechanical balance and box balance. The mechanical balance has a series of problems such as complicated structure, intricate design, high cost, scattered sensor, difficult debugging and installation, has been gradually replaced by a box balance based on the strain principle. In the verification stage, the test of the small thrust engine is mostly horizontal; the horizontal balance is generally used to directly measure the main thrust and the lateral force to reduce the measurement error of the main thrust and the lateral force of the horizontal plane. At the same time, in order to reduce the size limitation and processing difficulty, the aviation vector nozzle test platform adopts a modular box balance that can be processed separately by each component.

The six-component balance consists of three parts: three-dimensional force sensor, floating frame and fixed frame. The fixed frame mainly supports the main thrust and lateral forces. There are various mounting standards on the floating frame to ensure accurate positioning of the engine and accurate installation of each component. When the aeronautical vector nozzle test platform is working, the vector thrust is decomposed into three mutually orthogonal forces and three directions of torque by a six-component box balance. In order to properly match the three component force loads of the three-dimensional sensor and minimize the influence of the load mismatch of the balance components, it is necessary to adjust the layout between the four three-dimensional sensors reasonably. At the same time, in order to ensure the measurement accuracy and stability, the sensitivity of the sensor elastomer is required to be high, and the sensitivity of the sensor is required to be large. The crucial is the selection of elastomer materials and structural layout design [21].

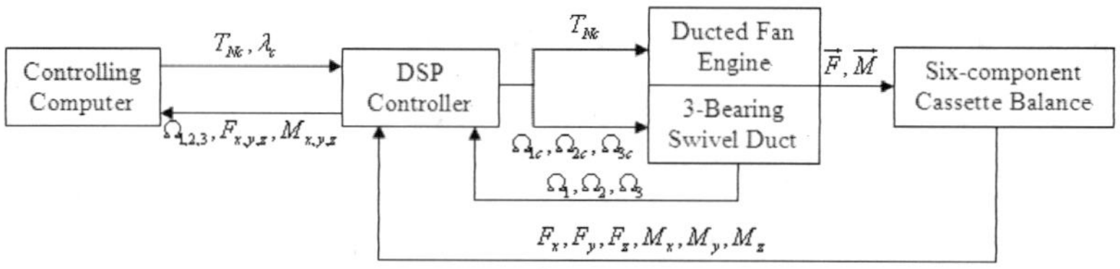

Figure. 19. Nozzle dynamics model

Figure. 20. Force test system

5. Conclusion

The difference between S/VTOL aircraft and conventional aircraft is mainly reflected in the field of power plant and aircraft-power plant depth coupling integration. These fields involve relatively new fundamental theoretical research and test technology. After decades of theoretical research, principle validation and engineering practice, the key technologies have been broken through, the test methods have been mastered, and the mature design and standard system have been established. Domestic research and principle experiments have been carried out in the field of S/VTOL aircraft power plant technology, but the research field is scattered and the technology maturity is low. The depth of research and the system integration are still difficult to meet the needs of development of short-distance take-off/vertical landing fighters. In general, power plant technology is still the bottleneck of China's development of S/VTOL aircraft. In order to speed up the development of S/VTOL aircraft power units, it is recommended that, on the basis of design and manufacturing of domestic aircraft and engine, materials technology, etc., a technical roadmap of S/VTOL aircraft power units must be , strengthen the aircraft demand and capability traction of engine, focus on the overall, core components and system design technology that restricts the power development, and realize the breakthrough of key technology integration.

6. References

[1] WANG X Y, ZHU J H, LIU K, ZHENG Y,Kinematic Modeling and Testing of 3-Bearing Swivel Duct Nozzle[J]. Acta Aeronautica et Astronautica Sinica 2014,35(4): 911-919

[2] Allen F J. Bolt upright: Convair's and Lockheed's VTOL fighters[J]. Air Enthusiast, 2007, 127: 13-20

[3] WANG Z X, LIU S, ZHOU L.Development and critical technology analysis on technology of short/vertical takeoff and landing fighter[J].Aeroengine, 2014, 40(4):1-6

[4] Bevilaqua P M. Joint strike fighter dual-cycle propulsion system [J]. Journal of Propulsion and Power, 2005, 21(5): 778-783

[5] Mason J R. F-35 S/VTOL Aircraft Performance Model Validation of Short Take-off Flight Test Data[R]. AIAA2013-4245

[6] Maddock I A,Hirschberg M J. The Quest for Stable JetBorne Vertical Lift:S/VTOL to F-35 S/VTOL[R]. AIAA2011-6999

[7] Saddington A J,Knowles K. Turbulence Measurement sin a Short Take-Off Vertical-Landing[J].Jouranl of Aircraft,2013,50 (2): 667-670

[8] Walker G P. Integrated Flight- Propulsion Control Development[R]. AIAA 2013-4243

[9] Jack Britt. Jet Pipe Arrangements For Jet Propulsion Engines[P]. US, 2933891, 1960

[10] Carroll E. Johnson. Variable Vectoring Nozzle for Jet Propulsion Engines[P]. US, 3260049, 1966

[11] Gerhard Kopp, Gero Otto Madelung. Swivelable Jet Nozzle, Intended Especially for Vertical Take-Off and Short Take-Off Planes[P]. US, 3443758, 1969

[12] Alexander Kurti, Richard W. Batt. Planetary Gear System Drive Mechanism[P]. US, 3485450, 1969

[13] Dudley O. Nash. Swivelable Jet Nozzle[P]. US, 3687374, 1972

[14] Michael C. Roberts. Apparatus for Moving Rotatable components. United States, B2, 7770379. Aug.10, 2010

[15] YUN Fei. Yak141-the first supersonic VSTOL fighter in world[J]. International Aviation,1991(12):28-30

[16] FANG Changde. Development research of aero-engine[M]. Beijing：Aviation Industry Press,2009：348-354

[17] Wang X Y,Zhu J H,Zhang Y J.Dynamics modeling and analysis of thrust-vectored S/VTOL aircraft[C]//The 32nd Chinese Control Conference,2013:1825-1830

[18] WANG X Y, YAN W H, XUE R R, ZHU J H and QI H.Design and Performance Analysis of a Scaled 3-Bearing Swivel Duct Nozzle[C] // The 2019 AIAA Propulsion and Energy Forum

[19] Ian A. Maddock, Michael J.Hirschberg. The Quest for Stable Jet Borne Vertical Lift:AS/VTOL to F-35 S/VTOL,AIAA 2011-6999

[20] HE W, TONG Z R, LI H B.Investigation of thrust valance for the single module scramjet[J].Journal of Aerospace Power.2010,25(10):2285-2289

[21] YU C A, LI P, WANG L.Design of elastic component of three-component strain gauge sensor[J].Chinese Journal of Sensors and Actuators,2012,25(3):333-337

IOP Conf. Series: Materials Science and Engineering **816** (2020) 012016 doi:10.1088/1757-899X/816/1/012016

Kinematics analysis and optimization of 6R manipulator

Qidong Li[1], Hehua Ju[1], Pengfei Xiao[1]

[1] Nanjing University of Aeronautics and Astronautics, Nanjing 210016, China

Email: juhehua@nuaa.edu.cn

Abstract. In order to solve the problem of 6R manipulator kinematics, this paper analyzes the inverse kinematics solution and optimizes the solution process of forward kinematics. Aimed to the structural characteristics of the manipulator, this paper first introduces the concept of the natural invariants, establishes the Denavit-Hatenber (D-H) model based on the natural invariants and proposes a new iterative algorithm for manipulator forward kinematics. Then, the inverse kinematics based on the natural invariants is proposed, which is divided into the position inverse solution and the inverse attitude solution, and the analytical solution is obtained by analyzing the singularity of the solution. Finally, the visual simulation model is built with MFC and Coin4D. The simulation results show that the proposed forward and inverse kinematics method of the six-degree-of-freedom manipulator has small error and high efficiency, and the accuracy and real-time performance of the algorithm are verified.

1. Introduction

The kinematics of manipulators is a central problem in the automatic control of manipulator. The kinematics of the manipulator includes positive kinematics and inverse kinematics. The positioning process of the manipulator is to calculate the joint variables of the manipulator according to the desired posture, so that the computer can control the manipulator. Domestic and foreign scholars have done a lot of research for the kinematics of the manipulator[1-4], at present, the analytic method[5,6] and the iterative method are mainly used to solve the inverse kinematics of the manipulator. The advantage of using the analytical method to solve the inverse kinematics of the manipulator is very obvious, all the solutions can be obtained by analyzing the model of the manipulator, and the solution is accurate and fast. Jorge Angeles[7] used Euclidean norm to solve the problem of six degrees of freedom decoupling manipulator, and simplified the solution of position inverse kinematics into a quartic equation of one variable by analytic method. The process of solving multivariate equations by iterative method is relatively simple, but it is impossible to determine how many sets of solutions are involved because of the initial value selection problem, and it is difficult to get all the solutions, and there is a danger that the iteration does not converge. Kalra[8] et al proposed an inverse kinematics algorithm for a six degrees of freedom manipulator based on genetic algorithm.

In this paper, by introducing the concept of natural invariants, a kinematic D-H model based on Axis Invariant is established and solved. On this basis, a new forward kinematics iterative algorithm is proposed. The inverse kinematics is divided into position inverse solution and inverse pose solution by pose separation and solving process. The analytical solution is obtained by analysis. Finally, the algorithm is verified correctness and feasibility.

2. Basic theory

According to the topological characteristics between the link of manipulator, the concept of the natural invariants is introduced. The movement of the manipulator can be more conveniently understood by the natural invariants.

2.1. Natural Invariants

When the system is at initial state, the coordinate system of axis $\bar{l}$ and axis l is $F^{[\bar{l}]}$ and $F^{[l]}$ respectively; the projection relationship between $F^{[\bar{l}]}$ and $F^{[l]}$ can be considered as linear transformation. Since both $F^{[\bar{l}]}$ and $F^{[l]}$ are Cartesian coordinate systems, an orthogonal base transformation matrix ${}^{\bar{l}}\boldsymbol{Q}_l$ can be used to describe:

$$\mathrm{A}\begin{pmatrix} e_{lx} & e_{ly} & e_{lz} \end{pmatrix} = \begin{pmatrix} e_{\bar{l}x} & e_{\bar{l}y} & e_{\bar{l}z} \end{pmatrix} {}^{\bar{l}}\boldsymbol{Q}_l \tag{1}$$

Taking the eigenvector corresponding to the eigenvalue of 1, and the eigenvector represented by the coordinate system of $F^{[l]}$ is ${}^{l}\boldsymbol{e}_{\bar{l}} = \begin{bmatrix} x_l & y_l & z_l \end{bmatrix}^{\mathrm{T}}$, which can be obtained:

$${}^{\bar{l}}\boldsymbol{Q}_l \cdot {}^{l}\boldsymbol{e}_{\bar{l}} = 1 \cdot {}^{l}\boldsymbol{e}_{\bar{l}} \tag{2}$$

Note is expressed as ${}^{\bar{l}}\boldsymbol{e}_l = \begin{bmatrix} x_{\bar{l}} & y_{\bar{l}} & z_{\bar{l}} \end{bmatrix}^{\mathrm{T}}$ in the $F^{[\bar{l}]}$ coordinate system. According to the linear transformation matrix ${}^{\bar{l}}\boldsymbol{Q}_l$, obtain

$$\begin{bmatrix} x_{\bar{l}} \\ y_{\bar{l}} \\ z_{\bar{l}} \end{bmatrix} = {}^{\bar{l}}\boldsymbol{Q}_l \begin{bmatrix} x_l \\ y_l \\ z_l \end{bmatrix} \triangleq {}^{\bar{l}}\boldsymbol{e}_l = {}^{\bar{l}}\boldsymbol{Q}_l \cdot {}^{l}\boldsymbol{e}_{\bar{l}} \tag{3}$$

Then the coordinate vectors represented by ${}^{l}\boldsymbol{e}_{\bar{l}}$ in $F^{[\bar{l}]}$ and $F^{[l]}$ coordinate systems are the same, namely:

$${}^{\bar{l}}\boldsymbol{e}_l = {}^{l}\boldsymbol{e}_{\bar{l}} \tag{4}$$

It can be seen from the above equation that ${}^{\bar{l}}\boldsymbol{e}_l$ and ${}^{l}\boldsymbol{e}_{\bar{l}}$ do not depend on the respective reference coordinate systems $F^{[\bar{l}]}$ and $F^{[l]}$, so ${}^{\bar{l}}\boldsymbol{e}_l$ or ${}^{l}\boldsymbol{e}_{\bar{l}}$ is said to be a natural invariant.

2.2. D-H Frame based on natural invariants

The D-H Frames plays a very important role in the kinematics calculation of the manipulator. The D-H system of each link is established based on the natural invariant. The D-H Frames of the link $\bar{l}$ and link l is represented by $F^{[\bar{l}]}$ and $F^{[l]}$ respectively, as shown in figure 1.

(i)Assume that $z_{\bar{l}}$ and z_l pass the natural invariants ${}^{\bar{\bar{l}}}\boldsymbol{e}_{\bar{l}}$ and ${}^{\bar{l}}\boldsymbol{e}_l$ respectively.

(ii)Draw the common perpendicular of axis $z_{\bar{l}}$ and axis z_l. Mark the points intersecting with $z_{\bar{l}}$ and z_l as $O_{\underline{l}}$ and O_l respectively, wherein $O_{\underline{l}}$ is the origin of the Middle Frame $F^{[\underline{l}]}$. O_l is the origin of Frame $F^{[l]}$ .Define the direction from axis $z_{\bar{l}}$ to axis z_l as axis x_l. If $z_{\bar{l}} \parallel z_l$, select one common perpendicular line to be axis x_l.

The D-H parameter is determined:

$c_{\bar{l}} = r_{\underline{l}}^{\bar{l}}$ - The Axial Offset Distance from the point $O_{\bar{l}}$ to the point $O_{\underline{l}}$.

$\phi_{\bar{l}} = \phi_{lx}^{\bar{l}x}$ - Rotational Angle from the axis $\bar{l}x$ to axis lx.

$a_{\bar{l}} = r_l^{\bar{l}}$ - The Axis Distance from the point $O_{\bar{l}}$ to the point O_l.

$\alpha_{\bar{l}} = \phi_{lz}^{\bar{l}z}$ - The Torsional Angle from the axis $\bar{l}_z$ to axis l_z.

${}_0\phi_{\bar{l}}$ - The initial angle between two adjacent x-axes.

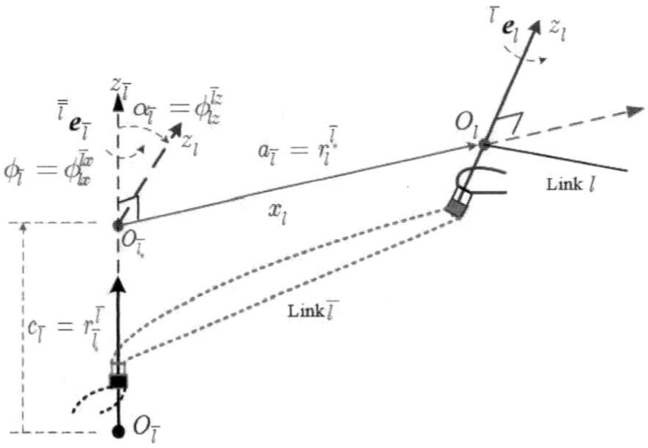

Figure 1. D-H Frame based on natural invariants.

From the above modeling process, the general formula of the transformation matrix of the adjacent rod coordinate system is: (Note: $C() = \cos(), S() = \sin()$)

$$ {}^{\bar{l}}\boldsymbol{Q}_l = \begin{bmatrix} C(\phi_{\bar{l}}) & -\lambda_{\bar{l}} \cdot S(\phi_{\bar{l}}) & \mu_{\bar{l}} \cdot S(\phi_{\bar{l}}) \\ S(\phi_{\bar{l}}) & \lambda_{\bar{l}} \cdot C(\phi_{\bar{l}}) & -\mu_{\bar{l}} \cdot C(\phi_{\bar{l}}) \\ 0 & \mu_{\bar{l}} & \lambda_{\bar{l}} \end{bmatrix} \tag{5} $$

Wherein $\lambda_i = C(\alpha_i)\, \mu_i = S(\alpha_i)$

$$ {}^{l}\boldsymbol{Q}_{\bar{l}} = {}^{l}\boldsymbol{Q}_{\bar{l}z} {}^{\bar{l}z}\boldsymbol{Q}_{\bar{l}} = {}^{\bar{l}}\boldsymbol{Q}_l^{-1} = {}^{\bar{l}}\boldsymbol{Q}_l^{\mathrm{T}} \tag{6} $$

The rotational transformation matrix ${}^{\bar{l}}\boldsymbol{Q}_l$ of the D-H system is a unit orthogonal matrix, and its inverse matrix ${}^{l}\boldsymbol{Q}_{\bar{l}}$ is equal to its own transpose. The characteristic of ${}^{\bar{l}}\boldsymbol{Q}_l$ provides great convenience for solving the inverse kinematics of manipulator.

3. Kinematics analysis of the manipulator

3.1. Forward kinematics of the manipulator

The schematic diagram of the industrial robot arm is shown in figure 2. From bottom to top, it consists of joints such as base, waist, shoulder, elbow, and wrist.

The kinematics model of the manipulator is obtained by obtaining the position and attitude of the end of the manipulator relative to the base of the manipulator. D-H Frame was established according to the kinematics model of the manipulator, D-H parameters are shown in table 1.

According to the D-H Frame of manipulator, the positive equation of motion is:

$$ \begin{gathered} {}^1\boldsymbol{r}_7 = {}^1\boldsymbol{r}_2 + {}^1\boldsymbol{Q}_2 \cdot {}^2\boldsymbol{r}_3 + {}^1\boldsymbol{Q}_2 \cdot {}^2\boldsymbol{Q}_3 \cdot {}^3\boldsymbol{r}_4 + \backslash \\ {}^1\boldsymbol{Q}_3 \cdot {}^3\boldsymbol{Q}_4 \cdot {}^4\boldsymbol{r}_5 + {}^1\boldsymbol{Q}_4 \cdot {}^4\boldsymbol{Q}_5 \cdot {}^5\boldsymbol{r}_6 + {}^1\boldsymbol{Q}_5 \cdot {}^5\boldsymbol{Q}_6 \cdot {}^6\boldsymbol{r}_7 \end{gathered} \tag{7} $$

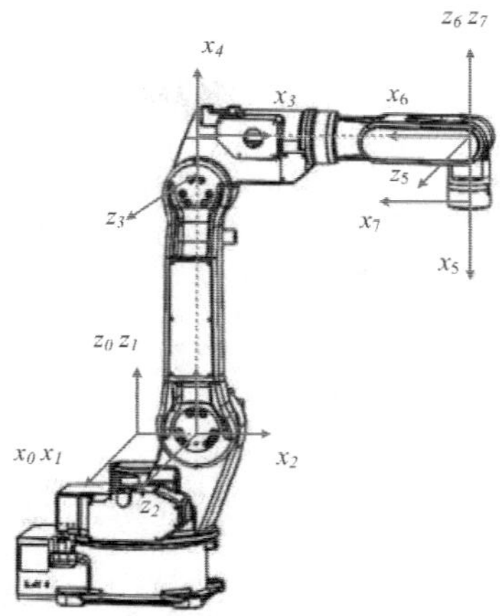

Figure 2. Schematic diagram of manipulator model.

Where ${}^{\bar{l}}\boldsymbol{r}_l$ is the coordinate of the origin of the coordinate system l in the $\bar{l}$ coordinate system.

Table 1. D-H parameters of manipulator.

	$a_{\bar{l}}$ (m)	$c_{\bar{l}}$ (m)	$\alpha_{\bar{l}}$ (deg)	${}_0\phi_{\bar{l}}$ (deg)
1	0.16	0.43	90	90
2	0.58	0	0	90
3	0.125	0	90	0
4	0	0.65	90	180
5	0	0	90	-90
6	0	0.16	0	0

Since positive kinematics involves a lot of matrix operations, it is very time consuming. In order to improve the speed of the positive movement of the manipulator, an iterative algorithm is proposed in this paper. The specific form is shown in (8).

$$ {}^1\boldsymbol{r}_7 = {}^1\boldsymbol{r}_2 + {}^1\boldsymbol{Q}_2 \cdot \begin{pmatrix} {}^2\boldsymbol{Q}_3 + {}^2\boldsymbol{Q}_3 \cdot ({}^3\boldsymbol{r}_4 + {}^3\boldsymbol{Q}_4 \cdot \\ \left({}^4\boldsymbol{Q}_5 + {}^4\boldsymbol{Q}_5 \cdot \left({}^5\boldsymbol{r}_6 + {}^5\boldsymbol{Q}_6 \cdot {}^6\boldsymbol{r}_7 \right)\right)) \end{pmatrix} \tag{8} $$

The iterative algorithm established in this paper avoids the matrix multiplication operation by calculating the product of the inner layer matrix and the three-dimensional vector. Compared with the positive motion equation established by (7), the calculation amount can be greatly reduced, as shown in table 2.

Table 2. Computational complexity of kinematic equations.

	Addition calculation	Multiplication calculation
General positive kinematics equation	117	153
Iterative positive kinematics equation	45	45

3.2. Position inverse solution

The position of the wrist of the orthogonal decoupling manipulator is only related to the first three joint angles, and the desired wrist position is ${}^1_d r_{4C}$.

$$ {}^1_d\boldsymbol{r}_{4C} = {}^1\boldsymbol{r}_2 + {}^1\boldsymbol{Q}_2 \cdot \left({}^2\boldsymbol{r}_3 + {}^2\boldsymbol{Q}_3 \cdot \left({}^3\boldsymbol{r}_4 + {}^3\boldsymbol{Q}_4 \cdot {}^4\boldsymbol{r}_{4C} \right)\right) \tag{9} $$

Let ${}^{l|\bar{T}}\boldsymbol{r}_l = \begin{bmatrix} a_{\bar{T}} & \mu_{\bar{T}} \cdot c_{\bar{T}} & \lambda_{\bar{T}} \cdot c_{\bar{T}} \end{bmatrix}^T$, then

$$ {}^{\bar{T}}\boldsymbol{r}_l = {}^{\bar{T}}\boldsymbol{Q}_l \, {}^{l|\bar{T}}\boldsymbol{r}_l \tag{10} $$

Where ${}^{l|\bar{T}}r_l$ represents the projection of position vector ${}^{\bar{T}}r_l$ in the l coordinate system.
Put (10) into (9), obtain:

$$ {}^{2}\boldsymbol{Q}_3 ({}^{3|2}\boldsymbol{r}_3 + {}^{3}\boldsymbol{Q}_4 \, {}^{4|3}\boldsymbol{r}_4 + {}^{3}\boldsymbol{Q}_5 \, {}^{5|4}\boldsymbol{r}_{4C}) = {}^{2}\boldsymbol{Q}_1 \left({}^{1}_{d}\boldsymbol{r}_C - {}^{1}\boldsymbol{r}_2 \right) \tag{11} $$

Take the Euclidean norm on both sides of the (11):

$$ A \cdot C(\phi_1) + B \cdot S(\phi_1) + C \cdot C(\phi_3) + B \cdot S(\phi_3) = -E \tag{12} $$

wherein：

$$ A = 2a_1 \cdot x_C \text{，} B = 2a_1 \cdot y_C \text{，} $$

$$ C = 2a_3 \cdot a_2 - 2c_4 \cdot c_2 \cdot \mu_3 \cdot \mu_2 \text{，} $$

$$ D = 2a_3 \cdot c_2 \cdot \mu_2 + 2c_4 \cdot a_2 \cdot \mu_3 $$

$$ \begin{aligned} E = {} & a_3{}^2 + c_3{}^2 + a_2{}^2 + c_2{}^2 + c_4{}^2 \\ & + 2c_3 \cdot c_2 \cdot \lambda_2 + 2c_4 \cdot c_3 \cdot \lambda_3 + 2c_4 \cdot c_2 \cdot \lambda_3 \cdot \lambda_2 \\ & + 2c_1 \cdot z_C - x_C{}^2 - y_C{}^2 - z_C{}^2 - a_1{}^2 - c_1{}^2 \end{aligned} $$

Since the third component of (11) is equal:

$$ F \cdot C(\phi_1) + G \cdot S(\phi_1) + H \cdot C(\phi_3) + I \cdot S(\phi_3) = -J \tag{13} $$

Wherein:

$$ F = y_C \cdot \mu_1 \text{，} G = -x_C \cdot \mu_1 \text{，} H = -c_4 \cdot \mu_2 \cdot \mu_3 \text{，} I = a_3 \cdot \mu_2 $$

$$ J = c_2 + c_1 \cdot \lambda_1 + c_4 \cdot \lambda_2 \cdot \lambda_4 + c_3 - z_C \cdot \lambda_1 $$

Express ϕ_1 by ϕ_3

$$ \begin{aligned} S(\phi_1) &= \frac{1}{\Delta_{11}} \begin{Bmatrix} F(-E - D \cdot S(\phi_3) - C \cdot C(\phi_3)) \\ - A(-J - I \cdot S(\phi_3) - H \cdot C(\phi_3)) \end{Bmatrix} \\ C(\phi_1) &= \frac{1}{\Delta_{11}} \begin{Bmatrix} G(-E - D \cdot S(\phi_3) - C \cdot C(\phi_3)) \\ - B(-J - I \cdot S(\phi_3) - H \cdot C(\phi_3)) \end{Bmatrix} \end{aligned} \tag{14} $$

Wherein $\Delta_{11} = -2a_1 \cdot (x_c^2 + y_c^2) \cdot \mu_1$。

Δ_{11} vanishes if and only if any of the factors a_1, $x_c^2 + y_c^2$ and μ_1 does. (12) or (13) contains only the unknown variable ϕ_3, and (15) will degenerate into a quadratic form. If $x_c^2 + y_c^2 = 0$, the wrist center is on the axis 1, and if a_1 、 μ_1 is simultaneously zero, axis 1 and axis 2 is coaxial and the configuration is wrong.

IOP Conf. Series: Materials Science and Engineering **816** (2020) 012016 doi:10.1088/1757-899X/816/1/012016

The squares of the two sides of (14) are respectively squared. And use the universal formula to replace $C(\phi_3), S(\phi_3)$ with $t = \tan\left(\phi_3/2\right)$.

$$R \cdot t^4 + S \cdot t^3 + T \cdot t^2 + Ut + V = 0 \tag{15}$$

Wherein:

$$\begin{aligned}
R &= 4a_1^{\,2}(H-J)^2 + \mu_1^{\,2}(C-E)^2 - 4a_1^{\,2}\mu_1^{\,2}\rho^2 \\
S &= 16a_1^{\,2}I(J-H) + 8\mu_1^{\,2}D(E-C) \\
T &= 8a_1^{\,2}(J^2+2I^2-H^2) + 2\mu_1^{\,2}(E^2+2D^2-C^2) - 8a_1^{\,2}\mu_1^{\,2}\rho^2 \\
U &= 16a_1^{\,2}I(J+H) + 8\mu_1^{\,2}D(E+C) \\
V &= 4a_1^{\,2}(H+J)^2 + \mu_1^{\,2}(C+E)^2 - 4a_1^{\,2}\mu_1^{\,2}\rho^2
\end{aligned}$$

The four sets of solutions obtained by (15) are solved, and ϕ_3 is further obtained.

$$\phi_3 = A\tan 2(\sin(\phi_3), \cos(\phi_3)) \tag{16}$$

Substituting ϕ_3 into (14) yields ϕ_1. ϕ_2 is calculated from the first two scalar equations of (11).

$$\begin{aligned}
C(\phi_2) &= \frac{1}{\Delta_{22}}\left\{\begin{matrix} X\cdot(x_c C(\phi_1) + y_c S(\phi_1) - a_1) \\ \backslash - Y\cdot\left(\begin{matrix}-\lambda_1 x_c S(\phi_1) + y_c\lambda_1 C(\phi_1) \\ \backslash + \mu_1\cdot(z_c - c_1)\end{matrix}\right)\end{matrix}\right\} \\
S(\phi_2) &= \frac{1}{\Delta_{22}}\left\{\begin{matrix} Y\cdot(x_c C(\phi_1) + y_c S(\phi_1) - a_1) \\ \backslash + X\cdot\left(\begin{matrix}-\lambda_1 x_c S(\phi_1) + y_c\lambda_1 C(\phi_1) \\ \backslash + \mu_1\cdot(z_c - c_1)\end{matrix}\right)\end{matrix}\right\}
\end{aligned} \tag{17}$$

Wherein:

$$\begin{aligned}
\Delta_{22} &= X^2 + Y^2 \\
X &= a_2 + c_4\mu_3\cdot S(\phi_3) + a_3\cdot C(\phi_3) \\
Y &= -a_3\lambda_2\cdot S(\phi_3) + c_4\lambda_2\mu_3\cdot C(\phi_3) + c_4\mu_2\lambda_3 + c_3\mu_2
\end{aligned}$$

If $\Delta_{22}=0$ (second singularity), then X, Y is zero at the same time, then the wrist is located on the axis of the second axis of rotation, then ϕ_2 can take any value, for convenience we can take a value of zero.

3.3. Attitude inverse solution

In this paper, the normal unit vector $\boldsymbol{N}$ and the radial unit vector $\boldsymbol{R}$ at the end of the manipulator are used to describe the attitude of the end of manipulator.

$$\begin{aligned}
\boldsymbol{N} &= {}^1\boldsymbol{Q}_6 \cdot {}_0\boldsymbol{N} \\
\boldsymbol{R} &= {}^1\boldsymbol{Q}_7 \cdot {}_0\boldsymbol{R}
\end{aligned} \tag{18}$$

Wherein ${}_0\boldsymbol{N}=\begin{bmatrix}0 & 0 & 1\end{bmatrix}^{\mathrm{T}}$, ${}_0\boldsymbol{R}=\begin{bmatrix}1 & 0 & 0\end{bmatrix}^{\mathrm{T}}$.

The last three axis of the orthogonally decoupled 6 DOF manipulator intersect at one point. The normal direction to the end of the manipulator is controlled by ϕ_4 and ϕ_5 to align with the end of desired normal direction, and the radial direction end of the manipulator is radially aligned with the end of desired radial direction by ϕ_6.

Given the normal direction ${}_d\boldsymbol{N}$ of the desired end of the manipulator, obtain:

$$ {}^4\boldsymbol{Q}_7 \cdot {}_0\boldsymbol{N} = {}^1\boldsymbol{Q}_4^{-1} \cdot {}_d\boldsymbol{N} \tag{19} $$

Put ${}_0N=\begin{bmatrix}0 & 0 & 1\end{bmatrix}^{\mathrm{T}}$ into (19), obtain

$$ \begin{bmatrix} C(\phi_4)\cdot C(\phi_5) \\ S(\phi_4)\cdot C(\phi_5) \\ -C(\phi_5) \end{bmatrix} = {}^{4|}_{d}\boldsymbol{N} \tag{20} $$

Wherein ${}^{4|}_{d}\boldsymbol{N} = {}^1\boldsymbol{Q}_4^{-1} \cdot {}_d\boldsymbol{N}$,

$$ \begin{cases} \phi_4 = A\tan\left({}^{4}_{d}\boldsymbol{N}^{[2]} \quad {}^{4}_{d}\boldsymbol{N}^{[1]}\right) \\ \phi_5 = A\cos\left({}^{4}_{d}\boldsymbol{N}^{[3]}\right) \end{cases} or \begin{cases} \phi_4{}' = -\phi_4{}' \\ \phi_5{}' = \phi_5 + \pi \end{cases} \tag{21} $$

After determining the joint angle ϕ_4 and ϕ_5, it is necessary to rotate the joint angle ϕ_6 to ensure that the radial direction of the end of the manipulator is aligned with the radial direction of the desired radial direction. Then

$$ \phi_6 = a\cos\left(\left({}^1\boldsymbol{Q}_6 \cdot {}_0\boldsymbol{R}\right)^{\mathrm{T}} \cdot {}_d\boldsymbol{R}\right) \tag{22} $$

It can be seen from (21) and (22) that there are at most 4 sets of attitude inverse solutions, and combined with the position inverse solution of the first three-axis, so the inverse solution of the 6 DOF manipulator has up to 16 groups.

4. Algorithm verification

In this paper, a set of visual robotic kinematics simulation tools is written by MFC and Coin4D, and the correctness of the inverse kinematics solution is verified by the method of forward and inverse kinematics mutual proof. Given the desired pose of the manipulator:

$$ \begin{aligned} {}^1_d\boldsymbol{r}_7 &= \begin{bmatrix}0 & \text{-0.215122} & 1.490120\end{bmatrix}^{\mathrm{T}} \\ {}_d\boldsymbol{N} &= \begin{bmatrix}0 & -1 & 0\end{bmatrix} \\ {}_d\boldsymbol{R} &= \begin{bmatrix}1 & 0 & 0\end{bmatrix} \end{aligned} \tag{23} $$

Deriving the desired wrist center coordinates according to the given position and posture, obtained by (24).

$$ {}^1_d\boldsymbol{r}_{4C} = {}^1_d\boldsymbol{r}_7 - 0.16 \cdot {}_d\boldsymbol{N} \tag{24} $$

The joint angles are solved by the simulation software written in this paper. The results are shown in table 3. The joint angles of the four sets are obtained by the inverse kinematics of manipulator (corresponding position inverse solution), the motion of the manipulator is shown in figure 3.

Table 3. Inverse kinematics results of the manipulator.

	$\phi_1(°)$	$\phi_2(°)$	$\phi_3(°)$	$\phi_4(°)$	$\phi_5(°)$	$\phi_6(°)$
1	0	45.000317	44.999490	0	0.000193	0
2	0	45.000317	44.999490	180	179.9998	180
3	0	8.566680	113.229455	180.000002	-148.203	180
4	0	8.566680	113.229455	0.000002	-31.7961	0
5	-180	20.158965	20.127770	-0.000003	-130.028	-180
6	-180	20.158965	20.127770	179.999997	-49.7132	0
7	-180	-43.100641	138.101175	0	174.9994	-180
8	-180	-43.100641	138.101175	-180	5.000534	0

Note: The solution of the inverse kinematics algorithm does not consider the two values of ϕ_6, so only 8 sets of solutions are obtained.

In order to verify the correctness and accuracy of the inverse kinematics of the manipulator, the output of the inverse kinematics is input into the forward kinematics, and the forward kinematics results are compared with the given expected poses to obtain the calculation error. As shown in table 4.

The obtained solution of inverse kinematics is substituted into forward kinematics, and the error is less than 10-7m， which proves the accuracy of inverse kinematics.

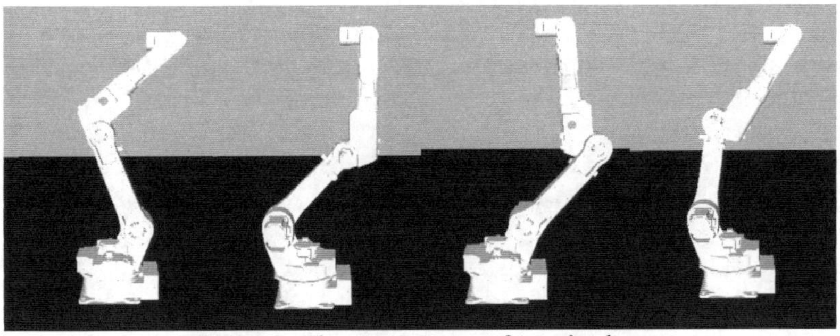

Figure 3. The movement of manipulator.

Table 4. Error of inverse kinematics results.

	Position error(mm)	normal vector error(mm)	radial vector error(mm)
1	7.95e-8	0	1.00e-10
2	7.95e-8	0	0
3	5.53e-8	2.30e-8	2.71e-8
4	5.52e-8	0	7.70e-9
5	5.33e-8	1.33e-7	0
6	5.33e-8	1.33e-7	2.39e-8
7	4.05e-8	1.33e-7	0
8	4.05e-8	1.33e-7	3.80e-9

Note: All error terms are the square root of the difference between the forward kinematic output and the given value. The zero mean that the error less than 1e-10.

5. Conclusion

In this paper, the motion model of the manipulator based on the natural invariants is established to solve the kinematics problem of the 6 DOF manipulator. A forward kinematic iterative algorithm is proposed, which is faster and more efficient than the traditional method of positive kinematics. At the same time, the method of pose separation is used to divide the solution process into position inverse solution and inverse pose solution, which reduces the complexity of the solution. Then the singular phenomena appearing in the solution process are analyzed and solved, and finally the analytical solution is obtained. Finally, the visual simulation of forward and inverse kinematics is carried out, which proves that the method is accurate, fast and efficient, and has important influence on improving the position accuracy and working efficiency of the manipulator.

Acknowledgement

This work was supported in part by the National Natural Science Foundation of China, grant number 61673010.

References

[1] T. Ao, M Liu, Z. Hua, Z. Hao 2017 Inverse Kinematic Solution of Industrial Robot Arm With Combination of Screw Theory and Algebraic Methods, *Mechanical Science & Technology for Aerospace Engineering*.

[2] H. H. Cheng 1994 Programming with dual numbers and its applications in mechanisms design, *Engineering with Computers*.**10** 212-229.

[3] S.S. Chiddarwar, N.R. Babu 2010 Comparison of RBF and MLP neural networks to solve inverse kinematic problem for 6R serial robot by a fusion approach, *Engineering Applications of Artificial Intelligence*. **23** 1083-1092.

[4] Y. Wei, S. Jian, H. Shuang, 2014 Z. Wang, General approach for inverse kinematics of nR robots. *Mechanism & Machine Theory*. **75** 97-106.

[5] S. Qiao, Q. Liao, S. Wei, H.J. Su, "Inverse kinematic analysis of the general 6R serial manipulators based on double quaternions. *Mechanism & Machine Theory*, 45, 193-199, 2010.

[6] V.J. Lumelsky 1984 Iterative Coordinate Transformation Procedure for One Class of Robots, *IEEE Transactions on Systems Man & Cybernetics*. **SMC-14** 500-505.

[7] C.M Gosselin 1997 Fundamentals of robotic mechanical systems: theory, methods, and algorithms, New York, (*International Journal of Robust & Nonlinear Control* vol 10) p1359-1359.

[8] P. Kalra, P.B. Mahapatra, D.K.Aggarwal 2006 An evolutionary approach for solving the multimodal inverse kinematics problem of industrial robots, *Mechanism & Machine Theory* **41** 1213-1229.

IOP Conf. Series: Materials Science and Engineering **816** (2020) 012017 doi:10.1088/1757-899X/816/1/012017

The Investigation of Parameters on the Microstructure of Friction Stir Welded 1060Al/Cu Butt Joint with 10mm Thickness

Y N Wei [1] *, H Li [1], F Sun [2], X Peng [1] and J T Zou [1]

[1]Department of Materials Science and Engineering, Xi'an University of Technology, 5 South Jinhua Road, Xi'an 710048, P.R. China
[2]Shaanxi Zhituo Solid-State Additive Manufacturing Technology Co., Ltd., 70 west of chaoyang street, Weinan 714000, P.R. China

Email: weiyanni@xaut.edu.cn

Abstract. T2 pure copper and 1060 aluminium alloy 10mm in thickness were joined by friction stir welding. The influence of fixed position and offset on the microstructure and hardness of joints was analysed. A joint having a good surface formation and a defect-free cross section was obtained when the copper was fixed on the advancing side and the offset was 3 mm. Layered structures, composite-like structures and clear copper side and nugget zone interface can be observed at the weld cross section. Under the combined action of heat and force, copper and aluminium inter-diffused and formed layered IMC at the interface. The layered IMC consists of Al_2Cu and Al_4Cu_9 by EDS and TEM analysis. The hardness of the joint first increases and then decreases from Cu side to Al side, and the hardness of the offset of 3mm is the lowest.

1. Introduction

Modern engineering structures often need to join heterogeneous materials, which are characterized by giving full play to their respective performance advantages and having good economic advantages [1-2]. Al/Cu composites can achieve high conductivity and thermal conductivity with a 40% reduction in mass and a price of only 60% of copper. Therefore, the joints of Al-Cu are often used in power transmission, refrigeration, and hydrometallurgy, and others fields, such as bus-bars, electrical connectors, heat-exchangers tubes, conductive tool, etc [3-5]. However, it is difficult to effectively weld Al and Cu using conventional fusion welding due to huge differences in their physical and chemical properties. And it is easy to form a large number of hard, brittle, high-resistance intermetallic compounds (IMCs) in the weld zone because of the strong chemical affinity between Al and Cu [6-7]. Thus, much attention has been directed towards solid state joining methods.

Solid phase method, such as explosion welding [8-9], diffusion welding [10-11], friction stir welding (FSW) [12], etc, has been used for welding Al to Cu in past decades. However, the explosion welding and the diffusion welding are mainly for the lap joint method, and the intermetallic compound layer at the interface is thick. FSW, an innovative solid-state welding technology invented by TWI, is a solid-state process which produces welds of high quality in difficult-to-weld materials such as Al and Cu [13]. In previous study, most of the research focused on the welding between Al and Cu plates below 6 mm. So far, there are few reports about the connection of Al and Cu thick plates. However, in practical industrial applications, joining of Al and Cu sheets of 10 mm or more are often used. In addition to the tool rotation rate and welding speed, the fixed location and the offset of the stir tip were the important parameters in the friction stir welding of Al/Cu dissimilar materials.

In this paper, the effect of fixed location of Al and Cu on the weldability of the joints was investigated and different offset was used to study the appropriate offset for friction stir welding of Al/Cu soft/hard dissimilar materials. The microstructure, phase composition, and bonding strength of the Al/Cu joints were analysed.

2. Experimental procedures

T2 pure Cu and 1060 Al plates 10mm in thickness, 200mm in length, and 70mm width was butt friction stir welded in this paper. The experimental tool adopted taper threaded tool pin profile which had a shoulder 25 mm in diameter and a pin 8 mm in diameter and 9.6 mm in length. The rotation speed is 950-1500 rpm, and the welding speed is 60 mm/min. In this paper, two fixed location of Al and Cu (Cu as the advancing side and Al as the advancing side) was used. The offset is the distance from the centre line of the stir tool to the butt line of Cu and Al. Three different sets of offsets were designed which were 1, 2 and 3mm respectively.

The welded joints were cut for section view, of which the structures and compositions were examined by optical microscope and scanning electron microscopy (SEM, JEOL, JSM-6700) equipped with an energy-dispersive spectrum (EDS). The hardness was carried out on a Vickers Hardness Tester.

3. Results and Discussion

3.1. The effect of fixed location on the weldability of the joints

The fixed location is one of the important parameters in the friction stir welding of Al/Cu dissimilar materials [14-15]. Figure 1 shows the macroscopic topography of the surface and cross-sectional at different fixed locations. The welding parameters are a rotation speed of 1500 rpm, a welding speed of 60 mm/min, and an offset of 3mm. As can be seen from Figure 1a and 1b, the weld surface is well formed, and a macroscopic defect-free joint is obtained when Cu plate is fixed on the advancing side. However, when Cu is fixed on the retreating side, the weld surface is poorly formed, and defects such as holes, tunnels, are observed. Figure 1c and 1d is the cross-sectional morphologies of the joint, Similarly, it can be seen from the cross section of the weld that Cu is placed on the advancing side without obvious defects. but when Cu is placed on the retreating side, a large void appears at the top of the weld, and the void extends in the welding direction to form Tunnel defect.

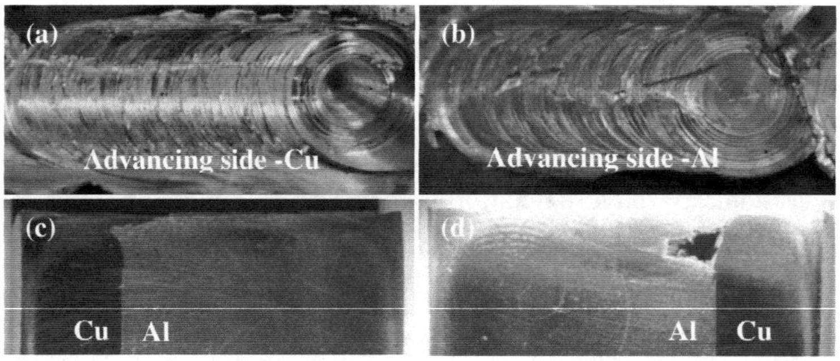

Figure 1. the surface and cross-section morphologies of the joints at different fixed location. (a)(c) Cu at the advancing side (b)(d) Al at the advancing side.

3.2. The effect of offset on the microstructure of the joints

In the similar materials butt friction stir welding, the tool pin is usually inserted into the centre of the joint line, at this time, the tool pin is 0mm offset. previous Studies shown that the quality of joints is poor at 0 mm offset, and it is difficult to obtain a joint without defects [14,16-17]. Therefore, in order to obtain a joint without defects, the tool pin is generally offset toward the softer material. So, in this experiment, the stirring needle is biased toward the Al side, and the offset amounts are 1 mm, 2 mm, and 3 mm, respectively. Figure 2 shows the effect of offset on the appearance and cross-section morphologies, it can be seen from Fig. 2a and b that a smaller offset cannot be obtained sound weld , At an offset of 1 mm, voids and cracks are observed on the weld surface and the cross section, and large-sized Cu blocks appear at the edges of the voids. When the offset is 2 mm, the weld surface is good, but there are hole defects inside the weld, the weld nugget zone is mixed with a large amount of Cu pieces and the Al matrix; and when the offset reaches 3 mm, the well-formed weld is obtained.

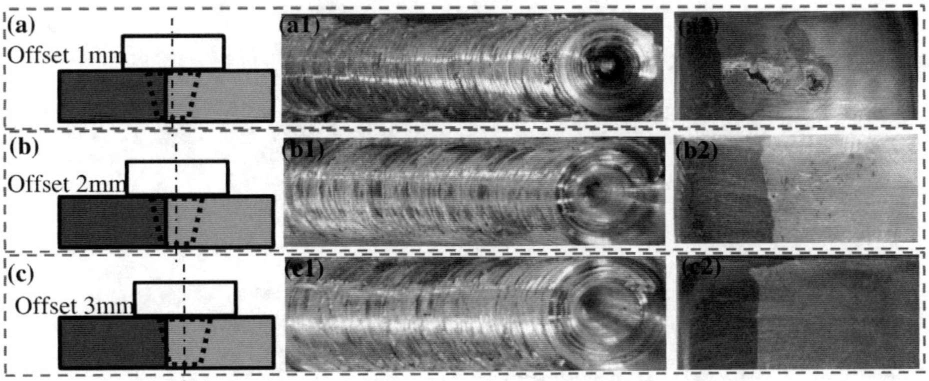

Figure 2. The surface and cross-section morphologies of the joints at different offset (a) 1mm, (b) 2mm, (c) 3mm

3.3. Microstructure of FSW joints

Figure 3 is a cross-sectional SEM image of the offset of 2 mm. As seen from Figure 3a and 3b, the weld nugget is formed by mixing a large number of different sizes of Cu fragments with Al matrix to formed a structure similar to an Al matrix composite. However, the presence of holes and cracks is observed near the Cu side at the bottom of the nugget zone. This may be due to the fact that the heat generated by friction stir welding is mainly provided by the shoulder, and the thickness of the sheet used in this paper is large, so the temperature gradient in the thickness direction is higher. The heat transfer and loss on the Cu side will be faster because the thermal conductivity of Cu itself is higher than that of Al. Therefore, the temperature near the Cu side at the bottom of the nugget zone is lower than that of other parts, and the material fluidity is worse, leading to the formation of unfused holes and cracks. On the contrary, as shown in Fig. 3c and 3d, the upper of the nugget zone generated more heat and the material fluidity is better, so that a good metallurgical bond is formed between Cu and Al without the occurrence of voids and cracks.

Figure 4 is a microstructure SEM of different regions. Figure 4a and 4b are the microstructure of the Cu-rich region at different offsets. When the offset is smaller, there are a small number of large-sized Cu blocks in the nugget zone, and the Cu-Al layered stack structure appears on the edges and inside of the Cu blocks, and light grey transition layers appeared in layers and layers (The colour is difference Cu and Al); With the increase of the offset, a large number of shredded and refined small-sized Cu pieces are uniformly dispersed in the Al matrix by the action of the thermo-mechanical coupling, to formed the structure similar to the Al-based composite material, and at the same time, the pieces reacted and generates IMC with Al matrix. By analysing the EDS of intermetallic compounds at different positions, it can be inferred that 1, 2, 3, and 4 points correspond to Al_4Cu_9, Al, Al+Al_2Cu, and Al_2Cu, respectively (i.e., Table1). Figure 4c and 4d shows that the Cu side and the nugget zone have a distinct interface, and a continuous, uniform IMCs with a thickness of about 1 μm was formed at the interface. This is because the Cu is fixed on the advancing side, and the material softened on the advancing side mostly flows forward along the welding direction, and the shear strain rate is larger. Secondly, due to the offset of the tool pin, the material in the nugget is mainly Al matrix. Therefore, the weld nugget zone forms a distinct interface with the Cu side.

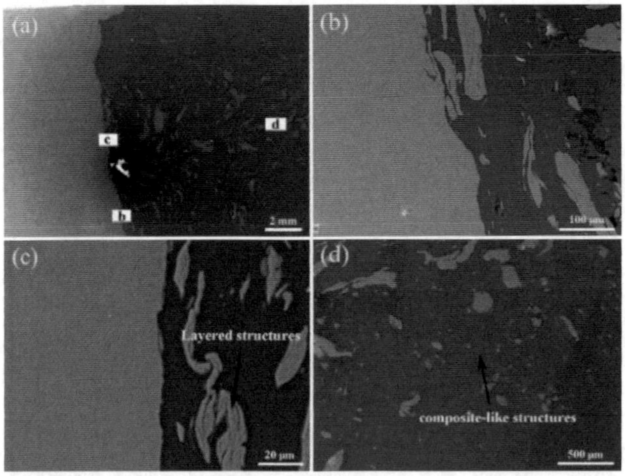

Figure 3. Cross-sectional structures of the Al-Cu joints.

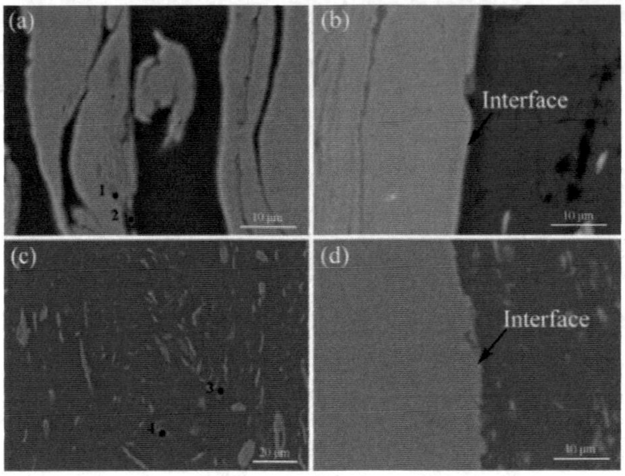

Figure 4. Microstructure of the Cu-rich and interface (a) (c) 2mm, (b) (d) 3mm.

Table 1 EDS results for different locations in figure 4.

Point	Al(wt%)	Cu(wt%)	Al(at%)	Cu(at%)	Possible phase
1	13.73	86.27	27.27	72.73	Al_4Cu_9
2	85.76	14.24	93.41	6.59	Al
3	72.16	27.84	84.45	15.55	$Al+Al_2Cu$
4	54.49	45.51	73.82	26.18	Al_2Cu

3.4. The Performance of joints

Due to the difference in hardness of the material itself, the micro-hardness of the joint shows an asymmetry along the centreline of the weld. Figure 5 is a profile showing the Vickers hardness of the weld cross-section at different offsets, it can be seen that the hardness trends first increases and then decreases from the Cu side to the Al side, the hardness value of the nugget fluctuates between the two base metals, and the maximum hardness value appears near the interface between the Cu side and the nugget. The nuggets hardness value with an offset of 3 mm is smaller than that with the other two offsets. This is due to the occurrence of river-like Cu-Al laminated structure similar to that in figure 4a in the structure of the nugget. Therefore, the hardness value varies with the laminated structure. When the offset is larger, the number of laminated structures in the nugget decreases, and correspondingly, the hardness value is lower. The contribution to the hardness of the nugget mainly depend on grain refinement.

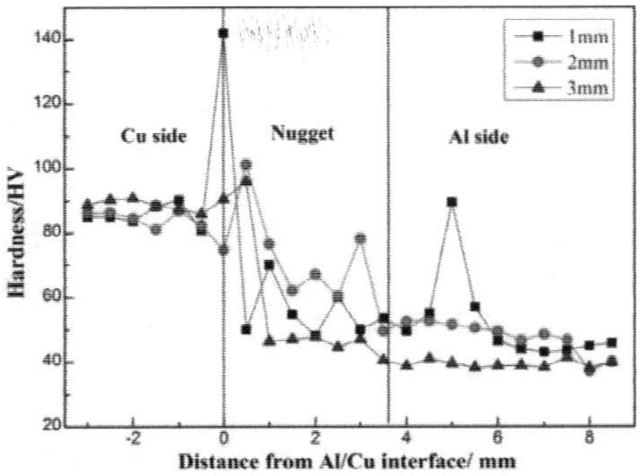

Fig 5. microhardness profile for Al–Cu joint at difference offset.

4. Conclusions

In this paper, the effects of fixed position and offset on the microstructure and hardness of FSW Al-Cu joints were discussed. The main conclusions are as follows:

(1) Al/Cu joint with good weld surface and no defect inside the cross section can be obtained when the Cu is fixed on the advancing side and the offset to the Al side is 3 mm.

(2) The Cu fragment which are peeled and shredded by the tool pin formed a layered structure and a composite structure in different regions of the joint. Atomic interdiffusion occurs at the interface between Cu and Al to form intermetallic compounds and these was identified as Al_2Cu and Al_4Cu_9.

(3) The trend of joint hardness is firstly increased and then decreased from Cu to Al, and the maximum value of hardness appears near the interface between the nugget and the Cu side.

References

[1] Huang G Q, Hou W T, Li J P and Shen Y F 2018 *Surf. Coat. Technol.* 344: 30–42.
[2] Xiong J, Peng Y, Zhang H, Li J and Zhang F 2018 *Vacuum* 147: 187–193.
[3] Khan H A, Wang K F and Li J J 2018 *Mater. Charact.* 141: 32–40.
[4] Liu J, Cao B and Yang J W 2017 *Metals* 7: 471.
[5] Tan C W, Jiang Z G, Li L Q, Chen Y B and Chen X Y 2013 *Mater. Des.* 51: 466–473.
[6] Mehta K P and Badheka V J 2017 *Trans. Nonferrous Met. Soc. China.* 27(1): 36–54.
[7] Mehta K P and Badheka V J 2015 *Int. J. Adv. Manufact. Technol.* 80: 2073–2082.
[8] Behcet G 2008 *Mater. Des.* 29(1): 275–278.
[9] Yuan Y, Chen P W, An E F and Feng 2018 *Mater. Sci. Forum.* 910: 52–57.
[10] Guo Y J, Liu G W, Jin H Y, Shi Z Q and Qiao G J 2011 *J Mater. Sci.* 46: 2467–2473.
[11] Eslami P, Karimi Taheri A and Zebardast M 2013 *J Mater. Eng. Perform.* 22: 3014–3023.
[12] Kumar R, Pancholi and Bharti R P 2018 *J. Mater. Process. Technol.* 255: 470–476.
[13] Mishra R S and Ma Z Y 2005 *Mater. Sci. Eng. R* 50: 1–78.
[14] Xue P, Ni D R, Wang D, Xiao B L and Ma Z Y 2011 *Mater. Sci. Eng. A* 528: 4683–4689.
[15] Debroy T and Bhadeshia H K D H 2010 *Sci. Technol. Weld. Join.* 15(4): 266–270.
[16] Singh S H and Mahmeen M 2016. Journal. Int. J. Mod. *Trends. Eng. Res.*3: 75–80.
[17] Esmaeili A, Besharati Givi M K and Zareie Rajani H R 2012 *Mater. Manufact. Process.* 27(12): 1402–1408.

Acknowledgments

This work is financially supported by the National Natural Science Foundation of China under Grant no. 51701154, the Key Research and Development Project of Shaanxi Province under Grant no. 2017ZDXM-GY-033 and 2018ZDXM-GY-136.

IOP Conf. Series: Materials Science and Engineering **816** (2020) 012018 doi:10.1088/1757-899X/816/1/012018

Fluid-structure Interaction Characteristics of Spin-on Fairing Structure of Aerodynamic Simulation Static Load

Longbin Liu[1, 2], Songyuan Yang[1] and Zhenyu Jiang[1]

[1]College of Aerospace Science and Engineering, National University of Defense Technology, Deya Road, Changsha, 410073, China

[2]Email: longbuaa@163.com

Abstract. With the rapid increase of aircraft speed, the impact of aerodynamic load on the aircraft is more significant, and aerodynamic load will cause the severe deformation of the structure, which will in turn affect the change of aerodynamic load. This paper takes a typical spin-on fairing structure as the study object with the finite element method, using simulated aerodynamic static load to study the effect of aerodynamic pressure on the load and deformation of the spin body structure, which could provide an effective method for the dynamic characteristics of the spin-on body structure, and for the study of aerodynamic load as well. And the results could also provide the initial finite element calculation data support for the damage evolution characteristics of fluid-solid coupling and long-term interaction of spin-on fairing structures.

1. Introduction

With the rapid improvement of aircraft flight performance, for the aircrafts such as missiles, airplanes, and rockets that need to interact with the atmosphere, because the working environment conditions of their flight are quite severe and complicated, especially among the high flying speed (up to supersonic speed or high sonic speed flight). The aircrafts usually need to bear the severe aerodynamic loads accompanied by aerodynamic heating problems, which results in significant changes in the load-bearing performance, flight performance and deformation characteristics of the aircraft structures [1-3], and in severe cases it may even cause the whole aircraft damage and failure with the aerodynamic load and structural deformation, which have attracted more and more attention [4,5]. Visbal and others [6-7] have studied the simplified boundary layer viscous fluid equation for subsequent research, which made a foundation pavement and a good start for subsequent in-depth research based on factors such as dynamic pressure, frequency, and Mach number. Liu, Kramer et al [8,9] studied the stiffness model for the unfolded and inflated flexible film inflatable tubes, and it was found that the unfolding of the folded inflatable tube was an unsteady and unstably fluid-structure coupling process, and the non-linear dynamic calculation model was also developed. Mohammad et al [10-12] studied the aerodynamic fluid-solid coupled dynamic expansion response of the thin film unit for the flexible film paraglider wing, constructed a cell model of the film unit, and combined with wind tunnel tests, and the the flow response was also expounded around the film and the structure characteristics. Wangwei et al [13-15] used fluent software to simulate the aerodynamic dynamics of the inflatable wing, and studied the finite element analysis method for simulating aerodynamic loads.

Based on the existing research results, taking the typical spin-on fairing structure as the object, the finite element method was used to study the effect of aerodynamic pressure on the bearing load and deformation of the structure by using simulated aerostatic load to obtain the simulated aerodynamic load on the spin-on structure. The influence law of deformation, and then the bearing deformation characteristics of different stress states and different deformation directions were obtained as well, which could provide an effective model for the comprehensive design and analysis of the thermal-fluid-structure coupling of aircraft structures, and provide the technology for the overall optimization design and performance analysis of missile fairings and flexible wings. Moreover, the results could provide the initial finite element calculation data for studying the fluid-solid coupling and damage evolution characteristics of the spin-on body structures.

2. The spinner fairing modelling

2.1. Aerodynamic load and structural deformation characteristics

The structure of the aircraft deformation usually occurs under the action of aerodynamic loads, and the bearing state of the fairing structure also changes accordingly. The finite element discrete method was used to study the model of the finite element mesh unit, and the displacement characteristics of any unit structure could be obtained.

$$\left\{u(x, y, z, t)\right\} = \left[N(x, y, z)\right]\left\{u^e(t)\right\} \tag{1}$$

Where, $\{u\}$ represents the displacement of any structure in the unit; $[N]$ representative shape function matrix; $\{u^e(t)\}$ is the node displacement vector, and the element strain and stress characteristics matrix of the structure deformation is obtained.

$$\{\varepsilon\} = [B]\{u^e\} \tag{2}$$

$$\{\sigma\} = [D]\{\varepsilon\} = [D][B]\{u^e\} \tag{3}$$

Where, $[B]$ is the geometric matrix, which represents the conversion relationship between displacement and strain; $[D]$ indicates the constitutive relationship matrix, which represents the conversion relationship between stress and strain.

For the aerodynamic load acting on the structured spinner fairing, the aerodynamic flow field satisfies the laws of conservation of mass, momentum, and energy. So the basic model for the description of the flow field parameters could be developed. And the air flow parameters field could be also obtained under the certain definite solution conditions. In order to simplify the finite element analysis model, the simulated aerodynamic static loads were used for the simulation calculations, and the simulated aerodynamic distributed pressure load was assumed directly acted on the spinner fairing structure to study the influence of aerodynamic pressure load on the structural load and deformation.

2.2. Finite element simulation model

The spinner fairing structure is susceptible to large aerodynamic loads. The influence of aerodynamic loads on the load and deformation of the head rectifier is mainly considered, and the corresponding local connection is simplified. The structure and dimensions of the fairing are shown in Figure 1, and the diameter is 2m, and the overall wall thickness is 7mm.

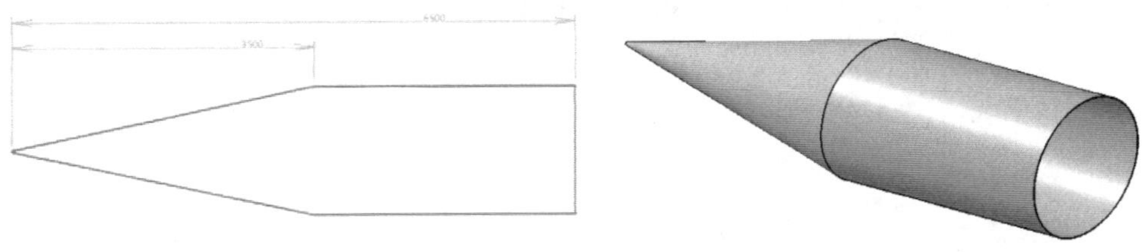

Figure 1. Schematic of fairing geometry model.

The fairing structure is generally made of natural or man-made dielectric materials, so that it could better adapt to maintain better heat resistance, wave permeability and higher structure strength in the harsher environment under the condition of meeting the aerodynamic shape, so as to protect the internal antenna or radar from the external environment, the new type of low-cost quartz ceramic composite material with good heat resistance was selected for simulation calculation, and the material parameters are shown in Table 1.

Table 1. The material parameters of spinning body fairing composite

Material name	density g/cm^3	Poisson's ratio	Elastic Modulus/ GPa
Composite material	2.2	0.15	48

The finite element software is used to divide the grid, and the structured grid is used to generate fast, good quality, and high calculation accuracy. Applying aerodynamic simulation loads, the forces under the actual working conditions are analysed and loaded on the hood to achieve simulation analysis. And the boundary condition is that the tail is fixed, and the front of the hood is loaded with aerodynamic pressure, which simulates the aerodynamic load at 3M, and the dynamic pressure is 577422Pa. Therefore, the above-mentioned load is loaded to the head cone, and the cylindrical surface of the fairing is not added with pneumatic Load simulation.

3. Coupling analysis of simulated aerodynamic static load on the fairing structure

As for the load bearing and deformation characteristics of the fairing, it coud be seen from the Mises stress distribution cloud diagram of the fairing that the Mises stress of the fairing has a circular distribution, which is due to its symmetrical properties. And the maximum stress is 73.85Mpa, which is close to the head, and the connection between the cover and the cylinder section is the minimum at the top of the head, which is almost 0, this is due to the small simulation aerodynamic load on the head, so the stress increases from the head to the tail direction, while the basic distribution of the segments is relatively uniform, which is about 40Mpa.

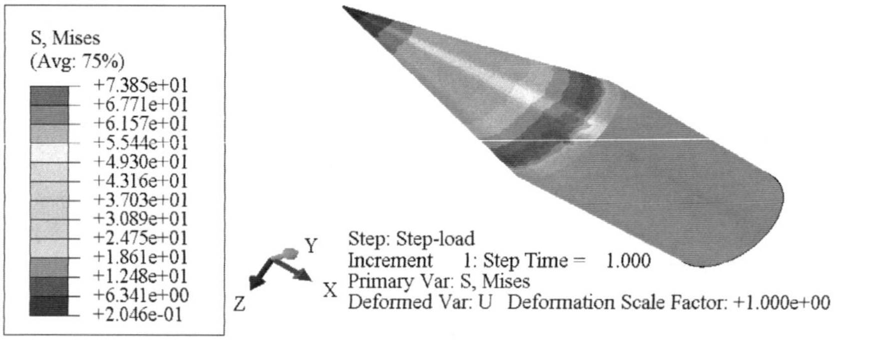

Figure 2. Cloud of the spine Mises stress distribution.

The compressive pressure of the spinning-on fairing structure presents a ring-shaped distribution, as shown in Figure 3, which is due to its symmetrical properties and loading with a circumferentially symmetrical load. The maximum compressive pressure is 55.43Mpa, which is close to the hood and the cylinder, and the connection point of the segment is the smallest near the vertex of the head. Therefore, the cylindrical segment from the warhead to the head cone has an enlarged form. Moreover, the cylindrical segment is basically evenly distributed, and most of them maintainst about 10Mpa.

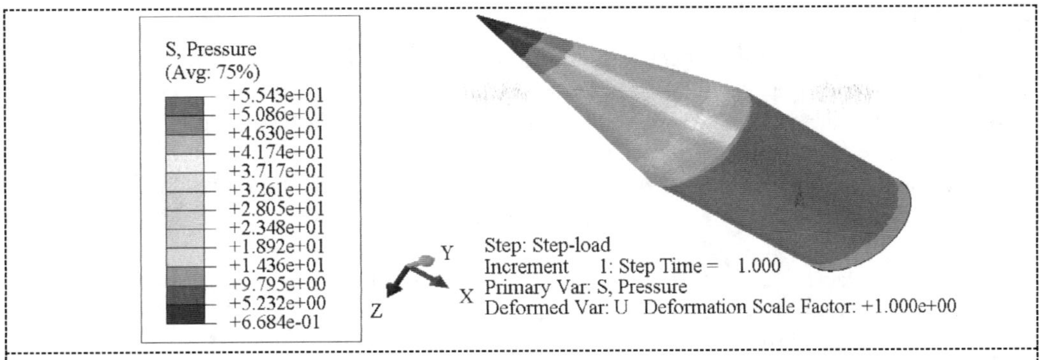

Figure 3. Cloud load distribution of spinner fairing pressure.

The displacement distribution of the whole fairing head cone could be seen in Figure 4, the displacement of the head is the largest (reaching 3.788mm). From the head to the cylindrical section, the displacement gradually decreases. And the whole displacement distribution also shows a ring, this is consistent with the symmetrical pressure distribution law of the rotating body. The displacement of the whole projectile head is around the range of 3.4-3.788mm, furthermore, the displacement of the rear section of the cylindrical section is gradually decreasing.

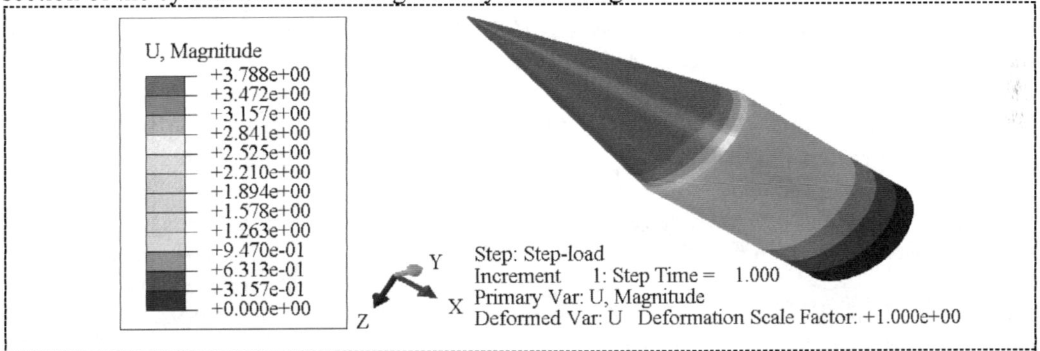

Figure 4. Cloud map of displacement of spinner fairing structure.

The maximum deformation strain cloud of the fairing head cone is shown in Figure 5, which shows a circular distribution law as a whole. The deformation strain of the head and the cylindrical section is relatively small, while the deformation strain of the rear part of the head cone is relatively large. So it could be seen that the deformation strain of the middle and rear parts of the head cone is large, and this area could be used as the key area for the optimal design.

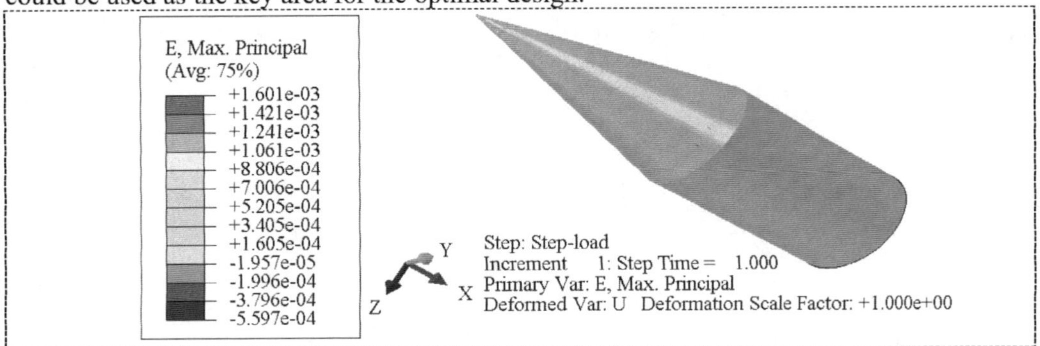

Figure 5. Cloud map of the maximum combined strain distribution of a spinner fairing.

4. Conclusions

(1) the Mises bearing stress of the fairing and the simulated aerodynamic pressure load are both distributed in the ring shape. And the maximum stress is close to the connection between the head cover and the cylindrical section, and the smallest is at the top of the head, and gradually increases along the axial direction, and the cylinder segment stress is more evenly distributed.

(2) The displacement distribution of the head cone of the fairing has the largest displacement of the head, which gradually decreases along the axial direction, and has a circular distribution, and the displacement of the rear section of the cylindrical section gradually decreases.

(3) The strain is also distributed in the ring shape, but the deformation strain of the head and the cylindrical section is relatively small, and the strain in the middle area is relatively large, so it is necessary to focus on the optimal design of this area.

References

[1] Wang W., Shen J 2006 *Air Missile* vol 8 pp 1-9. (in Chinese)
[2] Bowcutt K. G. 2001 *Journal of Propulsion and Power* vol 17(6) pp 1184-1190.
[3] Longatte E., Verreman V. 2009 *Journal of Fluids and Structures* vol 25(1) pp 95-111.
[4] Li X., Zhang J. W. 2014 *Aerospace Materials Technology* vol 1 pp 37-40, 56. (in Chinese)
[5] Ma W. X., Shen X. W., Lu X. Q., et al. 2005 *Applied Mechanics and Materials* pp 744-746.
[6] Visbal M, Gordnier R 2001 *15th AIAA Computational Fluid Dynamics Conference*.
[7] Xie C., Jia N., Shi D., et al. 2017 *Journal of Physics Conference Series* pp 916.
[8] Liu X. F., Tan H. F., Du X. W., et al. 2004 *Journal of Harbin Institute of Technology* vol 36(5) pp 684-687. (in chinese)
[9] Kramer R. M.， Cirak F.， Pantano C. 2013 *AIAA Journal* vol 51(7) pp 1640-1656.
[10] Mashud M., Umemura A. 2006 *Transactions of the Japan Society for Aeronautical and Space Science* vol 49(165) pp 154-161.
[11] Li Z., Fan X. L., Sun Q., et al. 2012 *Journal of Solid Rocket Technology* vol 35(5) pp 583-587.
[12] Chen J., Deng N., Li Y., e tal. 2017 *IOP Conference Series Earth and Environmental Science*, 61.
[13] Wang W., Wang H., Jia Q. P. 2010 *Hangkong Dongli Xuebao* vol 25(10) pp 2296-2301.
[14] Mir M., Colin H., Anthony C. Z., et al. 2014 *Acoustical Society of America Journal* vol 136(4) pp 2136-2136.
[15] Alexandr C., Elena F., Victor B., et al. 2018 *IOP Conference Series Materials Science and Engineering* vol 450 pp 022009.

Acknowledgement

This work was sponsored by the Hunan Natural Science Foundation"Hunan Province Science and Technology Plan Project Funding"(No. 2018JJ3591), and the National Natural Science Foundation of China (No.11802334), and National Basic Research Project (No.61XX010XXX03).

IOP Conf. Series: Materials Science and Engineering **816** (2020) 012019 doi:10.1088/1757-899X/816/1/012019

Numerical Study on the Effectiveness of Vectoring Primary Flow with Dilator by Using Microjet Actuator

Feng Gao, Qian Zhang*, Fengli Chen, Chengtao Zhang

Air and Missile Defense College, Air Force Engineering University, 710051 Shaanxi Xian, China

*Email: zhangqian8913@163.com

Abstract. In the present paper, orient control of dilator primary flow in microjet were simulated numerically by using microjet actuator with the X-L whole flow field calculation model. The effects of different expansion half angles of the primary flow channel, microjet actuator amplitude, micro jet actuator frequency, and the position of micro jet actuator and primary flow velocity on deflection of primary flow were analyzed. It is found that expansion half angle has great effect on the effectiveness of vectoring primary flow with dilator using microjet actuator; if the expansion half angle is small, the wall-attached flow of micro jet plays a main role in driving the primary flow flows along dilator wall; if the expansion half angle is big enough, the coupling effect between microjet and shear layer plays the main role. With the increase of the velocity of the actuator, the controlling efficiency of microjet is expanded. Besides, dilator decreases the distance sensibility between actuator and primary flow. With the increase of primary flow velocity, the controlling efficiency of microjet is decreased.

1. Introduction

In recent years, the zero-mass jet technology dominated by microjet technology has quickly become a hotspot in the research of flow control technology, and has received widespread attention from experts and scholars at home and abroad [1～9]. The experimental and theoretical research results show that the microjet as a control system is simple, light, consumes little energy and does not require a source of work, and has high control efficiency. It initially shows the advantages that other conventional jet control methods do not have. Application prospects.

Through experimental and numerical analysis, it is found that the nozzle with the dilator, the result of the interaction between the microjet and the main stream is very different from that of the non- dilator nozzle. And the expansion half angle of the dilator, the velocity amplitude of the microjet actuator, the frequency, the position of the microjet actuator and the main speed and other parameters have a great influence on the control efficiency of the mainstream. This paper intends to conduct a numerical analysis of this situation and to explore its mechanism.

2. Control Equations and Numerical Methods

The governing equations use a two-dimensional, incompressible Faver mean N-S equation, and the turbulence model uses the SST k-ω turbulence model. The microjet actuator uses a simplified model X-L full flow field computational domain model. The second-order upwind interpolation format is used to discretize the equation space, while the time discretization uses the first-order implicit format. The discrete solver is used to iteratively solve the discrete equations. See [3] for details.

IOP Conf. Series: Materials Science and Engineering **816** (2020) 012019 doi:10.1088/1757-899X/816/1/012019

3. Calculation Model and Initial Boundary Conditions

The calculation area and grid are shown in Figure 1, and the actuator structure dimensions are shown in Figure 2 and Table 1. In order to make the parameters on the boundary change very little, the numerical anomaly will not occur, and the external flow field is 300 mm × 300 mm. The mainstream channel width H is 12mm. The specific parameters of the main outlet are shown in Figure 2. The semi-expansion angle of the dilator is, the wall length L of the dilator is 6 mm, the actuator is placed on the wall of the dilator, and d is the central axis of the throat of the actuator and the left wall of the main channel. Distance, the central axis of the actuator is at the midpoint of the dilator wall, thus the distance of the microjet from the main flow. The actuator exit (flow center) and the wall boundary layer are mesh-encrypted.

Computational boundaries involve solid wall boundaries, free boundaries, and inlet boundaries. As shown in Figures 1 and 2, B1 is a solid wall boundary with a non-slip boundary condition, and the wall's pressure gradient and temperature gradient are zero. The B2 mobile exit is taken as a free boundary. B4 is the speed entry boundary given by the actuator simplified model X-L.

The microjet is established in a stationary ambient fluid, so the initial conditions of the flow field calculation are zero except for the B4 inlet boundary, the pressure is normal pressure, and the temperature is normal temperature. In this case, the actuator frequency f is taken as 400 Hz and the amplitude is taken as 2 mm.

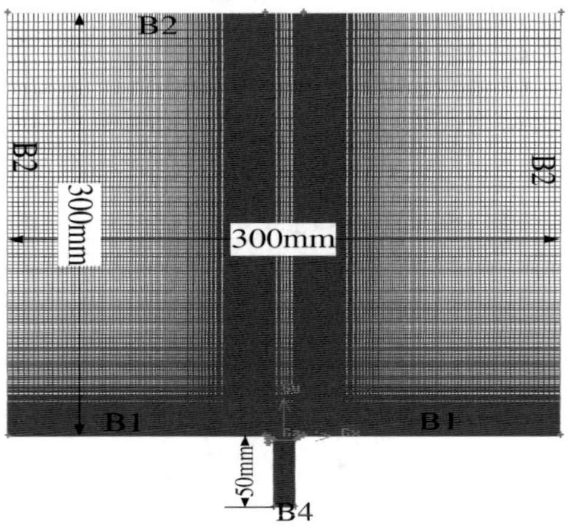

Figure 1. Computational domain and grid.

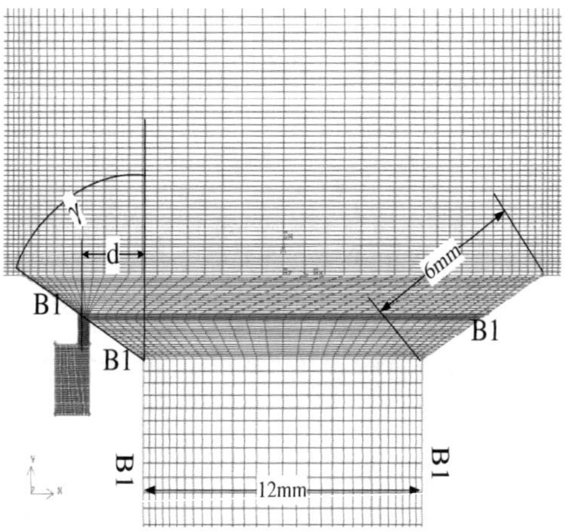

Figure 2. Partial view of the grid at the main exit.

4. Calculation results and analysis

4.1 Calculation conditions

The calculation conditions of the microjet actuator for the main control of the belt dilator are shown in Table 1. The mainstream speed is the velocity amplitude of the microjet vibrating film. Case0 is the basic working condition when the semi-expansion angle of the dilator is 20° and the actuator is not working. Case1 to Case8 are numerical experiments in which the microjet actuator is controlled in the mainstream when the semi-expansion angle of the dilator is different. When the microjet actuator frequency f is 400 Hz and the midpoint amplitude of the actuator film is 2 mm, the maximum velocity of the midpoint of the actuator film is $U_m = 5m/s$.

Table 1. Microjet actuator mainstream vector control calculation conditions.

Working condition	$\gamma(°)$	A_m(mm)	f(Hz)	U_m(m/s)	d(mm)	U_0(m/s)
Case0	20	0	0	0	1.0261	5
Case1	10	2	400	5	0.5209	5
Case2	20	2	400	5	1.0261	5
Case3	30	2	400	5	1.5	5
Case4	40	2	400	5	1.9284	5
Case5	50	2	400	5	2.2981	5
Case6	60	2	400	5	2.5981	5
Case7	70	2	400	5	2.8191	5
Case8	80	2	400	5	2.9544	5
Case9	50	0.4	400	1	2.2981	5
Case10	50	1.2	400	3	2.2981	5
Case11	30	4	200	5	1.5	5
Case12	30	1	800	5	1.5	5
Case13	60	4	200	5	2.5981	5
Case14	60	1	800	5	2.5981	5
Case15	70	2	400	5	1.5	5
Case16	70	2	400	5	5	5
Case17	70	2	400	5	2.5981	10
Case18	70	2	400	5	2.5981	15
Case19	40	2	400	5	1.9284	10
Case20	40	2	400	5	1.9284	15

4.2 The effect of expanding half-angle γ on the mainstream

It can be seen from Fig. 3 that the effect of different γ angles on the main flow of microjets is very obvious. In the case of the microjet actuator power and frequency, the microjet deflects the main stream to 40.77° at γ =30°, and the microjet increases the deflection angle φ of the main stream as the angle A of the dilator increases. Consistent with the Pack and Seifert experiments, the microjet achieved a maximum deflection angle of 52.63° when γ =50° was reached. When the angle is further increased, the γ =70° microjet deflects the main stream to 28.50°, and the microjet reduces the deflection angle of the main stream to not only increase but also decrease. Fig. 4(a) is a pressure diagram of the wall surface of the main channel under different A (L represents the left wall surface and R represents the right wall surface). For comparison, the formula $y^* = y + 0.006\cos(\gamma)$ is defined. It can be seen from Fig. 4(a) and Fig. 4(b) that when γ =50°, the low pressure region generated between the microjet and the main flow has the greatest influence on the pressure gradient of the main channel. This is because the distance between the microjet and the main stream is small when γ =30°, and the distance between the microjet and the main channel is larger in the y direction, which is not conducive to the upstream propagation of the low voltage region along the main channel. When γ = 70°, the distance between the microjet and the main stream is too large, which is not conducive to the establishment of the low pressure zone. At the same time, the microjet actuator controls the main deflection efficiency. On the other hand, when the γ angle is less than 50°, the lower half of the dilator restricts the deflection of the main stream.

When the γ angle is greater than 50°, As the angle A increases, the dilator provides sufficient space for the main deflection, but the wall attachment effect of the flow decreases, while the distance between the microjet and the main stream increases, the influence of the microjet on the main stream decreases, the low pressure region decreases, and the low pressure dominates the mainstream. The influence of the channel is reduced, so the control efficiency of the microjet to the mainstream is reduced.

In summary, when the γ angle is less than 50°, the lower half of the dilator restricts the deflection of the main flow; while γ is greater than 50°, the wall attachment effect of the microjet flow is weakened, and the distance between the microjet and the main stream is increased, and the low pressure is low. The influence of the zone on the mainstream channel is weakened, and the control efficiency of the microjet to the mainstream is reduced. Therefore, $\gamma = 50°$ is the optimum value for the semi-expansion angle in the selected operating conditions.

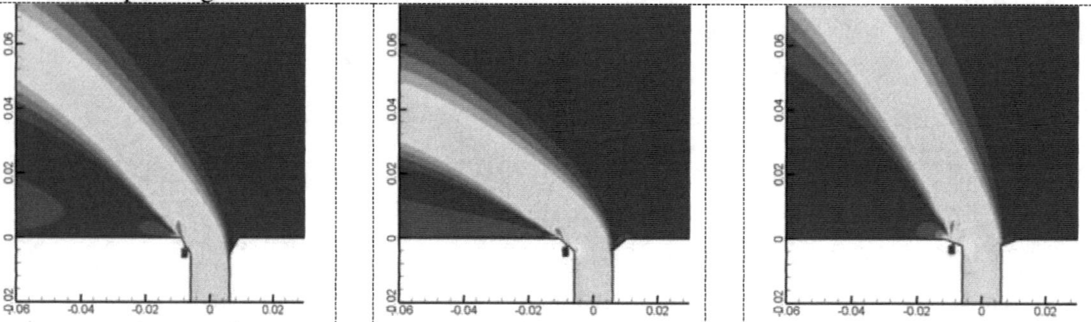

Case3（φ=40.77°） Case5（φ=52.63°） Case7（φ=28.50°）

Figure 3. Flow field velocity contour map of mainstream vector control of different γ angle microJet actuators.

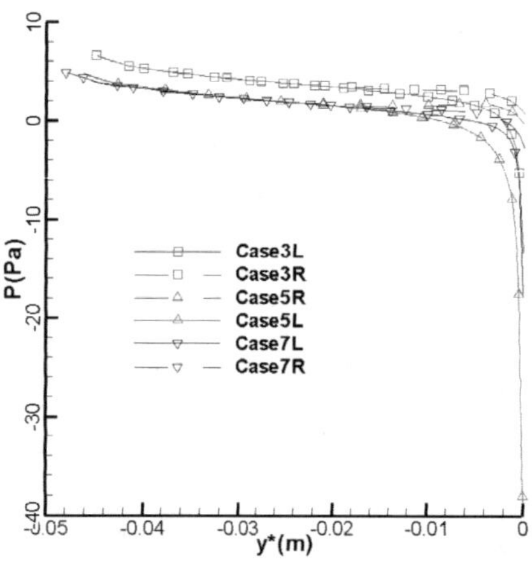

Figure 4(a). Comparison of the pressure of the left and right wall surfaces of different γ angles.

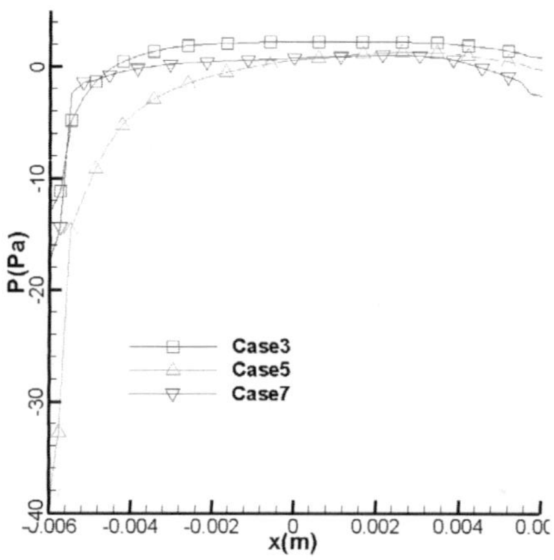

Figure 4(b). Comparison of lateral pressure at the joint at different γ angles.

4.3 Effect of actuator amplitude U_m on the mainstream

Case5, Case9, and Case10 simulate the effects of different speed amplitudes of microjet actuators on mainstream control. Table 2 shows that when $U_m = 5\text{m/s}$, $U_m = 1\text{m/s}$, and $U_m = 3\text{m/s}$, the frequency f is 400 Hz, the mainstream speed U_0 is 5 m/s, and the deflection angle of the main stream when the γ angle of the dilator is 50°. It can be seen from Table 2 that the larger the velocity amplitude, the higher the control efficiency of the microjet to the mainstream. On the one hand, the larger the velocity amplitude of the microjet actuator, the more fluid the microjet takes between the microjet and the main stream in the " expiration " process due to the shearing force; the inhalation of the microjet and the mainstream in the "sucking" process The more fluid there is, the stronger the low-pressure zone formed between the microjet and the main stream, and the greater the influence of the low-pressure zone on the upstream channel during the upstream propagation process. As shown in Fig. 5(a), the Case5 mainstream channel can be seen. The pressure difference between the left and right walls is the largest, and the lateral pressure gradient at the junction of the main channel and the dilator is also the largest in Fig. 5(b). On the other hand, when the amplitude of the microjet actuator is reduced, the vorticity of the vortex pair generated by the microjet is small (Fig. 6), and the vorticity of the microjet is difficult to break the free shear layer generated by the mainstream, thus controlling the mainstream. It is weakened. In summary, the larger the amplitude of the microjet actuator, the higher the control efficiency of the microjet to the mainstream.

Table 2. Effect of different amplitudes of actuators on mainstream deflection.

Working condition	Case5	Case9	Case10
Deflection angle φ (°)	52.63	6.218	35.50

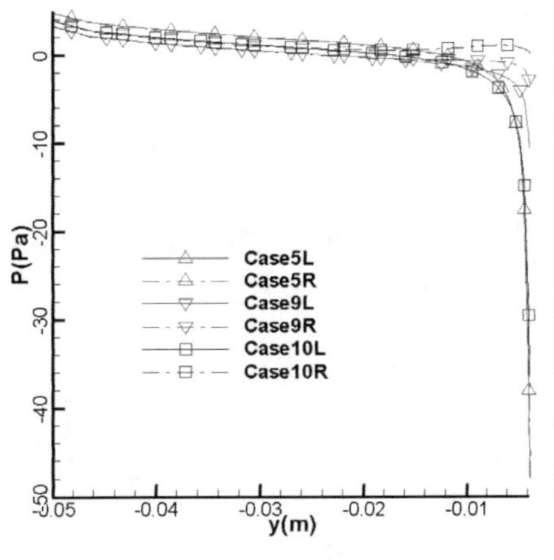

Figure 5(a). Comparison of left and right wall pressures of different amplitudes.

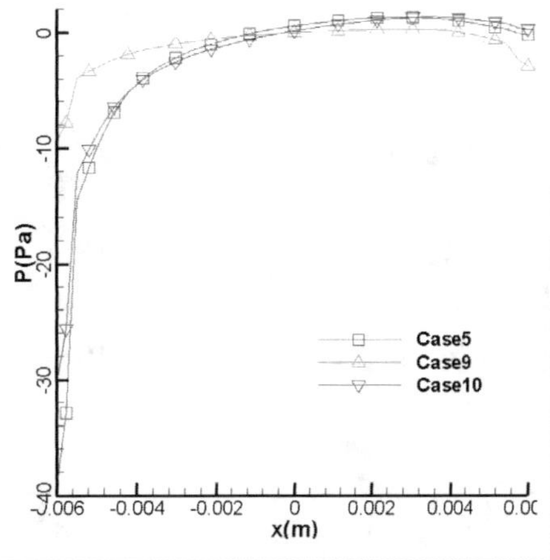

Figure 5(b). Comparison of lateral pressure at different amplitude joints.

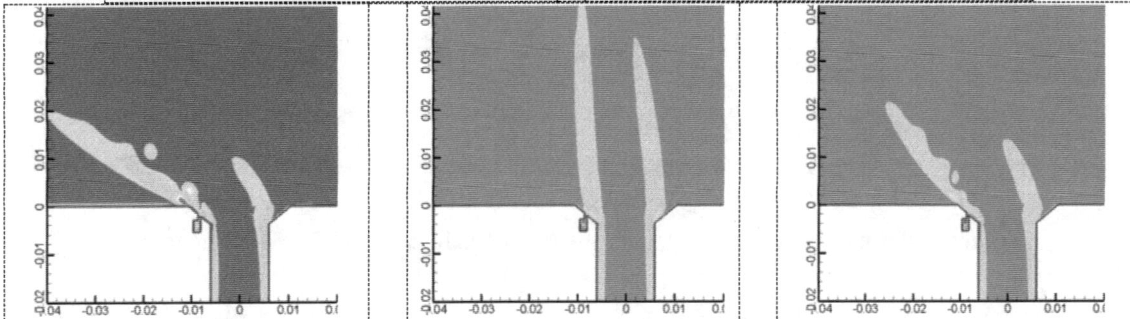

Case5 Case9 Case10

Figure 6. Comparison of different amplitude microJet vortices.

4.4 *Frequency f impact on the mainstream*

When the γ angle of the Case3, Case11, and Case12 simulating dilator is 30°, the effect of the microjet actuator on the mainstream control at different frequencies; When the γ angle of the Case6, Case13, and Case14 simulating dilator is 30°, the effect of the microjet actuator on the mainstream control at different frequenciesTable 3 shows the mainstream control efficiency of the microjets in different conditions of Case3, Case6, and Case11- Case14. It can be seen from Table 3 that when the γ angle of the divergence angle is 30°, the change of the frequency of the microjet actuator does not change significantly to the main control effect, which is consistent with the Pack and Seifert experiments; and when the γ angle is 60°, The control efficiency of the microjet to the mainstream is quite obvious.
As can be seen from Fig. 7, the frequency of the microjet actuator is increased, and the vorticity of the microjet is greatly increased near the main exit, However, when the AAA angle is 30°, the AAA angle is relatively small, and the vortex of the microjet does not have sufficient space to be fully developed, so that the vortex pair can only migrate downstream along the semi-expansion angle of the dilator. Therefore, the dilator limits the control of the microjet to the mainstream.

Table 3. Influence of microjets on the mainstream at different frequencies.

Working condition	Case3	Case11	Case12	Case6	Case13	Case14
Deflection angle φ (°)	40.77	40.53	40.80	35.12	34.27	41.14

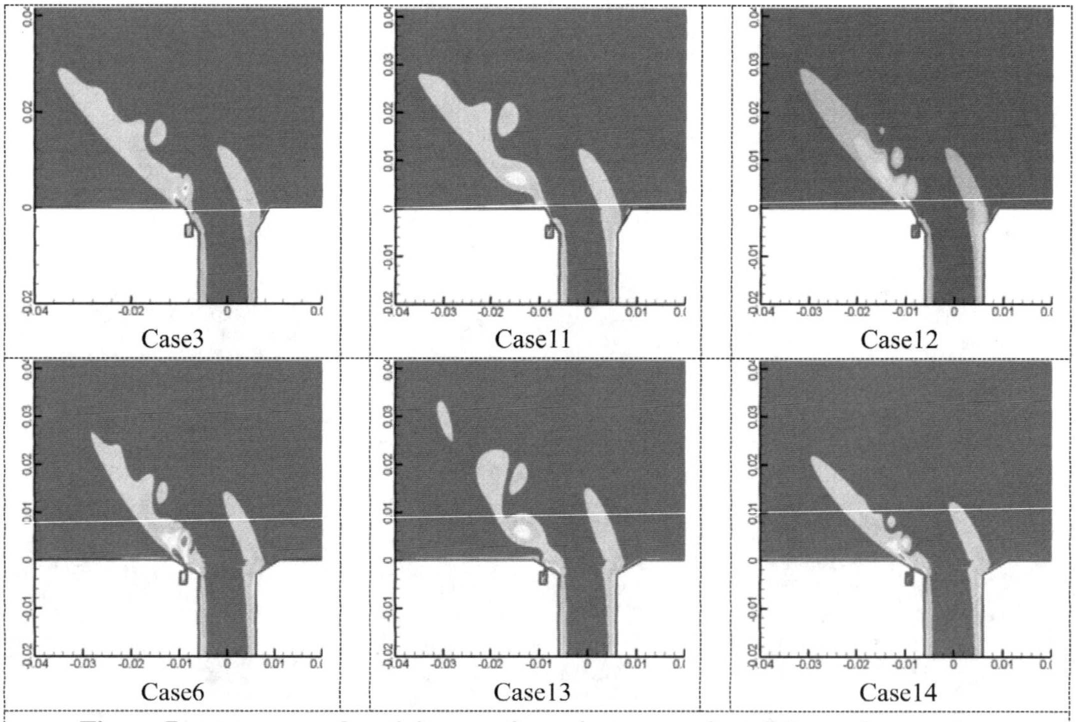

Figure 7. ontour map of vorticity near the mainstream exit at different frequencies.

When the AAA angle reaches 60°, the wall effect of the microjet is not obvious when it is 50°. But the microjet vortex has plenty of room to develop. Therefore, as the frequency of the microJet actuator increases, the vortex vorticity increases. The increase of the microjet free shear layer significantly enhances the traction of the mainstream. Therefore, the effect of the microjet on the main flow control is enhanced.

4.5 Effect of actuator position

Table 4 shows the effect of the microjet actuator on the main flow control at different positions of the dilator when the γ angle of the dilator is 70°. As can be seen from Table 4, when the microjet actuator is in the middle of the dilator, that is, the microjet actuator is at a distance of $L\sin(\gamma)/2=6\sin(70°)/2=2.8mm$ from the main flow, the microJet has the best control efficiency to the main stream, and the actuator is located in other parts of the dilator. That is, the distance between the microjet and the mainstream $c\sin(\gamma)=1.6\sin(70°)$ = 1.5mm or 5mm is not as good as Case7. This is because the distance between the Case15 microjet and the mainstream is too small to establish a low-pressure recirculation zone, which causes the pressure difference between the left and right wall surfaces to affect the control efficiency. Case16 is due to the increase of the distance between the microjet and the main stream, and the effect of the microjet free shear layer on the main stream is reduced to reduce the efficiency of the microjet to the mainstream control. From the selected working conditions, the optimal distance of the microJet to the main control is 2.8mm (5.6h), not 3h without the dilator [4,5]. Therefore the dilator changes the optimal distance of the microjet to the mainstream control. At the same time, it can be seen from Table 4 that the main deflection angle does not change much, without the jet vector control of the dilator, the mainstream is very sensitive to d, and the mainstream is almost unaffected by the actuator when d>5h [4, 5], that is, the dilator reduces the sensitivity of the distance between the actuator and the main stream. In Fig. 8, when the semi-expansion angle is 70°, the velocity cloud map of the mainstream is controlled by microjet at

Table 4. Control effect of microjets at different positions on the mainstream at an angle of 70°.

Working condition	Case7	Case15	Case16
Deflection angle φ（°）	28.13	25.74	27.32

In summary, the dilator increases the optimal distance of the microjet to the mainstream control; at the same time reduces the sensitivity of the microJet control efficiency to d, which reduces the placement accuracy of the microjet actuator in the flow field.

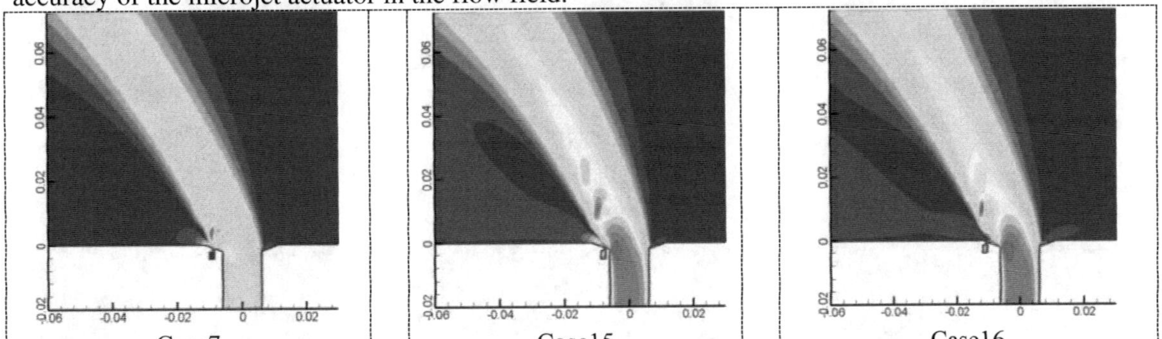

Figure 8. The velocity cloud diagram of the mainstream of microjet control at different positions of γ.

4.6 Effect of mainstream speed

Table 5 shows the control effect of the microjet on the mainstream under different mainstream speed conditions when the γ angle of the dilator is 70°. It can be seen from Table 5 that as the mainstream speed increases, the control efficiency of the microjet to the mainstream decreases. This is because when the mainstream speed increases, the fluid absorbed by the actuator from the main flow to the cavity is relatively less in the "suction" process, and the free shear layer formed by the main stream is less likely to be broken, and the control efficiency of the microjet to the mainstream is reduce. Therefore, if the mainstream speed is too large, in order to obtain a better deflection effect, it is also necessary to increase the amplitude of the actuator to obtain a large exit speed. Figure 9 is a flow field velocity diagram of the mainstream of different speeds when the γ angle is 70°. Table 6 shows the effect of microjets on the control efficiency of the mainstream under different mainstream speeds when the γ angle is 40°. It can be seen from Table 6 that when the γ angle is 40°, the control effect of the microjet has changed significantly with the change of the mainstream speed, but the control effect of the microjet on the mainstream is better than that when the γ angle is 70°. The efficiency is high. Therefore, when the mainstream speed is increased, the dilator suitable for the semi-expansion angle can be used to control the main stream. Figure 10 is a flow field velocity diagram of the mainstream of different speeds when the γ angle is equal to 40°.

Table 5. Control effect of microjet on different speed mainstream when γ angle is 70°.

Working condition	Case7	Case17	Case18
Deflection angle φ (°)	28.13	15.58	9.64

Table 6. Control effect of microjet on different speed mainstream when γ angle is 40°.

Working condition	Case4	Case19	Case20
Deflection angle φ (°)	49.44	33.69	21.76

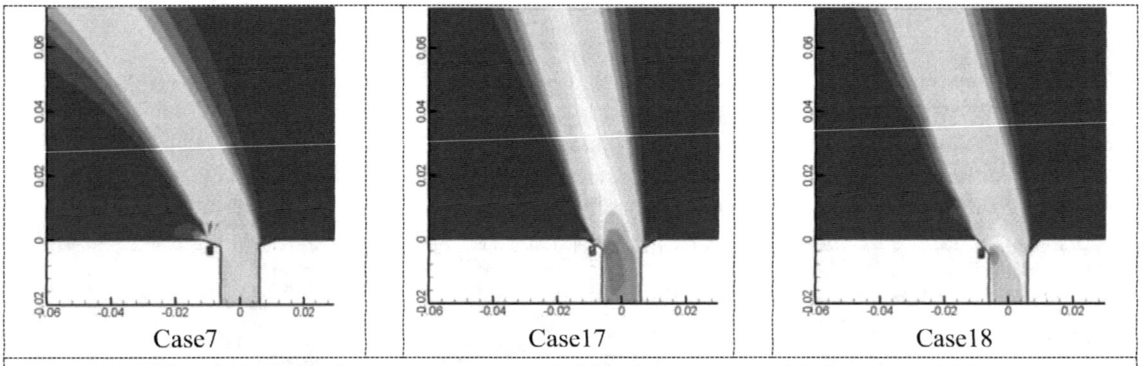

Figure 9. Flow field velocity map of different speeds at AAA angle of 70°.

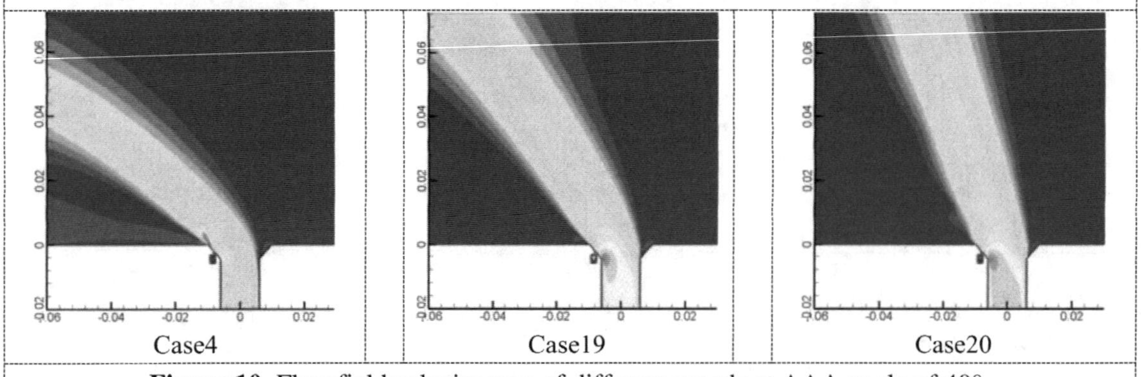

Figure 10. Flow field velocity map of different speeds at AAA angle of 40°.

5. Conclusion

In this paper, the full-flow field calculation model X-L model of the microjet actuator is used to simulate the mainstream of the microjet directional control belt expander. The calculation results show:

1. The semi-expansion angle of the dilator has a significant effect on controlling the main stream. When the AAA angle is small (AAA <50°), the wall-attached flow of the microjet plays a major role, which makes the mainstream almost along The wall of the dilator faces downstream; when the angle is large (AAA > 50°), the coupling of the microjet and the main shear layer plays a major role.
2. The dilator can reduce the sensitivity of the frequency to the microjet vector control effect, so that the microjet vector control reduces the requirement on the microjet frequency characteristics.
3. The dilator can reduce the sensitivity of the distance between the microjet actuator and the main stream to the microJet control effect, thereby reducing the precision of the arrangement of the microjet actuator in the flow field.
4. In the case of a large mainstream speed, a suitable expander can be selected to implement the most effective control of the

6. References

[1] Mcmicawl J M. Progress and Prospects for Active Flow Control Using Microfabricated Electro–Mechanical System (MEMS) [R]. AIAA 96-0306.

[2] Smith B L and Glezer A, Vectoring and Small-Scale Motions Effected in Free Shear Flows Using Synthetic Jet Actuators [R]. AIAA 96-0213.

[3] Lin Peiji. Microjet and its application in jet vector control [D]. Master's thesis of Air Force Engineering University, 2009.

[4] Gao Feng, Wang Liang. Mainstream Numerical Simulation of Directional Control of Single Micro Jet Actuator[J]. Journal of Northwestern Polytechnical University, 2003, 21(6): 528～531.

[5] Gao Feng, Wang Liang, Zhang Zhifeng. Analysis of Main Parameters of Directional Control of Single Micro Jet Actuator[J], Journal of Air Force Engineering University, 2003, 4(2): 12~15

[6] Gao Feng, Wang Liang. Numerical Simulation of External Flow Field of Double Micro Jet Actuator[J]. Journal of Aerodynamics, 2003, 21(3): 267~274.

[7] Rathnsing ham R, Breuer K S. Coupled fluid-structural characteristics of actuators for flow control [J]. AIAA Journa1,1997, 35 (5).

[8] Chen F J. Development of synthetic jet actuators f or active flow control at NASA Langley[R]. A I AA 2000, 2405.

[9] Pack L G, Seifert A. Periodic excitation for jet vectoring and enhanced spreading [J]. Journal of Aircraft, 2001,38(3):486-495.

IOP Conf. Series: Materials Science and Engineering **816** (2020) 012020 doi:10.1088/1757-899X/816/1/012020

Star Sensor Layout Design Based on PSO

ZHA Micheng, ZHAO Liping

Technology and Engineering Center for Space Utilization, Chinese Academy of Sciences, System design laboratory, Beijing, 100094

zhamicheng17@mails.ucas.ac.cn

Abstract. Incident light, such as sunlight and atmosphere-earth light, will interfere with the viewing field of star sensor, and the accuracy and precision of star sensor will decrease. A design method based on PSO algorithm and STK software to solve layout problems efficiently is applied. The star sensor layout problem is described as the problem of finding the global optimal solution in three-dimensional space. The results show that this method can solve the problem of satellite sensor layout quickly and effectively and the method has certain engineering application value.

1. Introduction

Star sensors are high-accuracy optical sensors. In order to capture more targets, a large field of view angle needs to be designed. However, it should try to avoid the incident light that affects the precision and accuracy of star point extraction, such as sunlight and atmosphere-earth light, entering the star sensor field of view. The design of the hood structure and the incident light filtering algorithm can improve the anti-incident ability. Optimizing the design of the star sensor installation layout position and the direction of the field axis also reduces incident light effects. In order to ensure the effectiveness of the star sensor, the installation layout of the star sensor should be analysed in detail in the design stage of the whole star scheme [1].

An et al. took the incident light suppression performance of the star sensor as the research focus and they used STK to analyse the angle change between the optical axis of the star sensor and the solar vector in a typical time [2]. Du et al. provided the conditions for incident light to enter the viewing field of the star sensor. They proposed an analysis method for the viewing field of star sensors based on STK and they simulated incident light entering the viewing field [3]. Zhang et al. established a mathematical model for the incident light in the viewing field of the star sensor. They used STK to simulate the field of view of incident light entering the star sensor, and gave the time period, duration and other information [4]. On the basis of studying the geometric position of Sun-Earth-Lune-Star, Zhao Lin et al. analysed the influence of sunlight on the view of star sensor during the process of satellite on orbit operation by combining the installation position of the star sensor and the viewing field [5].

The above several types of the viewing field analysis are only based on STK software, and the layout position needs to be designed repeatedly and iteratively. It is difficult to find the optimal layout quickly and the efficiency is low. In order to avoid the influence of incident light, the orientation of the satellite sensor installed on the satellite was analysed and optimized to ensure the effectiveness of the satellite sensor in orbit. Particle swarm optimization algorithm(PSO) is a population-based stochastic optimization technique. In this paper, the conditions of incident light entering the viewing field of star sensor are analysed and the viewing field analysis model of star sensor is established. Combined with the data provide by STK tool and PSO algorithm, the validity of the layout method is verified by simulation.

2. Coordinate System Establishment

In this paper, geocentric inertial coordinate system J2000 is used to describe the position relationship among earth, satellite and satellite orbit and to determine the satellite position and velocity parameters. The stability of the satellite attitude usually adopts three modes: three-axis stability, spin stabilization, and gravity gradient stabilization. In this paper, the three-axis attitude stabilization method is selected, which is suitable for satellites with various pointing requirements. The zero-attitude schematic diagram of the satellite and the ontological coordinate system are shown in figure 1. The center of mass is the origin O_e of the coordinate system. The X-axis is constrained by the direction of inertial velocity, and the Z-axis always points to the Nadir.

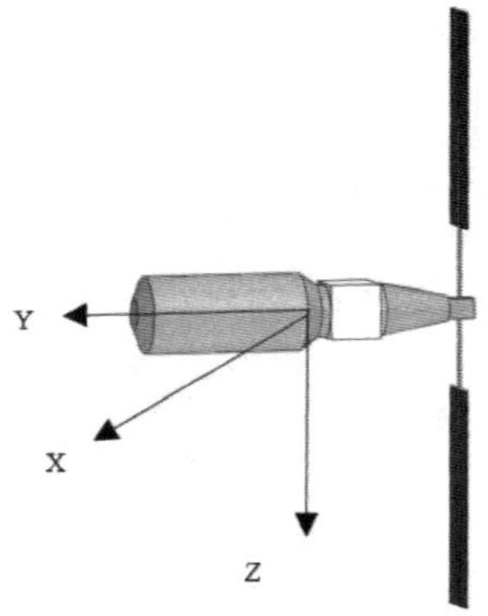

Figure 1. Satellite body coordinate system.

3. Analysis of incident light entering the viewing field

Incident light which makes the background noise of CCD camera increase, enters the viewing field of star sensor. Incident light reduces the level of the entire image surface and deteriorates the sharpness. In severe cases, the target image or signal may be completely overwhelmed by incident noise. Incident light will seriously affect the accuracy and reliability of star point extraction [6].

The suppression angles of sunlight and ground-gas light in the star sensor are γ_s、γ_m、γ_e. S_S is the sun direction vector in the star sensor coordinate system. The angle between the vector S_S and the boresight direction of star sensor is β_{SS}, as shown in figure 2.

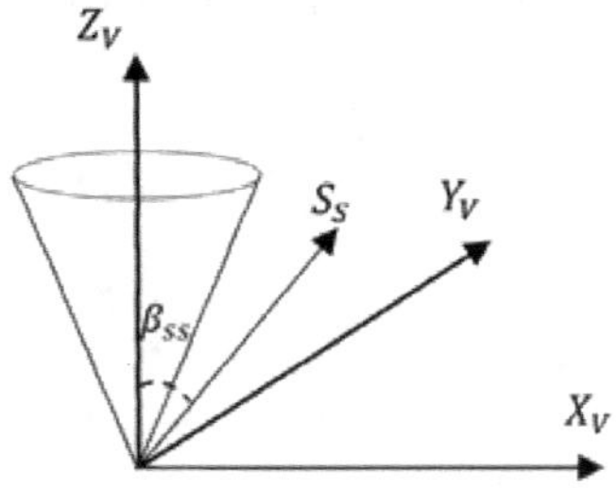

Figure 2. Angle relationship between the sun vector and the visual axis vector of the star sensor.

3.1. Conditions for sunlight to enter the viewing field

Due to the distance between the earth and the sun, the solar diffusion angle observed from the earth is only about 30 ', so sunlight can be thought of as parallel light. The angle between the sun direction vector and the satellite position vector is α_{SS}. The radius of the earth is R_e. The tangent line between the sun's rays and the earth intersects with the satellite's flight orbit at points L1 and L2, as shown in figure 3.

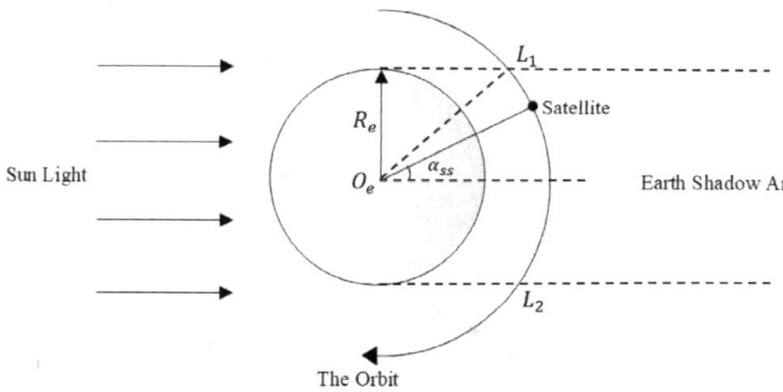

Figure 3. Location relationship analysis of sunlight-earth-satellite

According to figure 4, it can be analysed:

(1) When the satellite enters the ground shadow area, i.e. $\alpha_{ss} < arcsin\,(\frac{R_e}{|O_e L_1|})$, there is no direct sunlight in the viewing field, so this region is not considered.

(2) When $\alpha_{ss} \geq 90°$, the satellite is illuminated by the sun. Moreover, when $\beta_{ss} \leq \gamma_s$, sunlight enters the viewing field of the star sensor.

(3) When $90°\text{-arcsin}\,(\frac{R_e}{|O_e L_1|}) \leq \alpha_{ss} < 90°$ and $\beta_{ss} \leq \gamma_s$, sunlight into the viewing field.

3.2. Conditions for the entry of atmosphere-earth light into the viewing field

The earth's surface is covered with an atmosphere more than 1000 kilometers thick. The atmospheric system is a space system with the upper surface of the earth as the lower limit and the upper boundary of the atmosphere as the upper limit. The earth can be regarded as a blackbody with a temperature of 300K. And its radiation peak wavelength is about 10 μm [7]. In the study of the influence of atmosphere-earth light on the irradiance of star sensor, the irradiance of the reflected sunlight on the earth's surface is mainly studied, regardless of the influence of the earth's own radiation. Due to the reflection and scattering of sunlight from the atmosphere, the atmosphere-earth light becomes a strong incident light source which affects the normal operation of the star sensor. The atmosphere-earth light seriously interfered with the working process of the satellite sensor on the LEO orbit satellite [8].

When the satellite enters the sunlit area, the atmosphere-earth light will cause interference. When the satellite is in the shadow area, there is no atmosphere-earth light interference. Therefore, only the satellite in the solar exposure area was analyzed. Suppose that the earth's atmospheric system is spherical, the atmospheric thickness is d_a and the satellite's center of mass is O_L. The atmosphere-earth light exists in the cone with satellite centroid as the apex, $O_L A$ as element, $O_L O_e$ as central axis and α_{se} as semi-cone Angle. The angle between the atmosphere-earth light and the visual axis of the star sensor is β_{se}, which is the angle between the conical surface and the visual axis outside the conical surface, as shown in figure. 4.

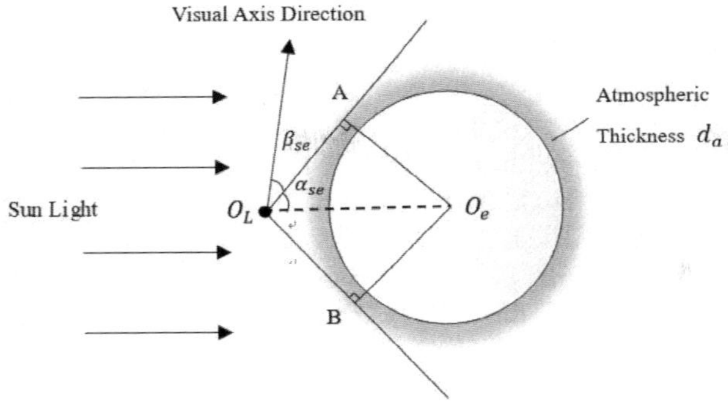

Figure 4. Atmosphere-earth light into the field of view analysis

According to the earth shadow model in figure 5, the conditions for atmosphere-earth light to enter the viewing field of the star sensor are as follows:

(1) When $\alpha_{ss} \geq arcsin\,(\frac{R_e}{|O_e L_1|})$, the satellite is under the sun light. And when the visual axis of the star sensor is within the semi-cone angle α_{se}, the atmosphere-earth light enters the viewing field;

(2) When $\alpha_{ss} \geq arcsin\,(\frac{R_e}{|O_e L_1|})$, the satellite is under the sun light. And when $\beta_{se} \leq \gamma_e$, the atmosphere-earth light enters the viewing field.

3.3. Comprehensive conditions of incident light entering the field of view of star sensor

Based on the above single incident light analysis, the specific process of determining whether the field of view of the star sensor is exposed to atmosphere-earth light and sunlight is shown in figure 5.

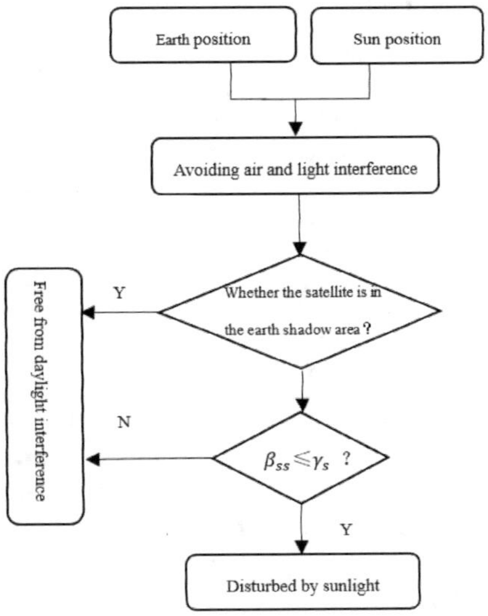

Figure 5. Flow field of incident light entering star sensor

According to this flow chart, it is specifically determined whether the viewing field of the star sensor is affected by stray light. During the orbiting of the satellite, the direction of the boresight only needs to satisfy $\beta_{se} \geq \gamma_e$ to avoid the atmosphere-earth light. When the visual axis direction of the star sensor satisfies the conditions of $\beta_{ss} \geq \gamma_s$, sunlight and moonlight can be prevented from entering the field of view.

4. Simulation model establishment

In the previous chapter, the influence of incident light on the different positions of star sensor is analysed in detail, and the conditional model of incident light entering the star sensor is described. In this chapter, the satellite toolkit STK and MATLAB are used to analyse the influence of incident light on the viewing field of the star sensor. Developed by analytical graphics, STK is a leading business analysis software in the aerospace field. It supports the entire process of space missions, including design, test, launch, operation and mission applications, and provides analytical results in the form of diagrams and texts.

4.1. Add target satellite model

This article starts by creating a scenario in STK. The angle between the earth-sun line and the orbit plane in a year is the smallest on the summer solstice. At this time, the Angle between the star sensor and the sunlight is the smallest, which has the greatest influence on the star sensor. Therefore, only the worst case is considered in the simulation modelling [8]. Therefore, the day with the smallest solar incidence angle in a year was selected, and the simulation time is set as 21 Jun 00:000.000 UTCG -22 Jul 00:000.000 UTCG.

Use the "Orbit Wizard" method to add a target satellite S_c with a circular orbital height of 400 km and an orbital inclination of 42°. Ideally, the satellite attitude angle is (0°, 0°, 0°). The pose type is the direction of the nadir and the direction of the velocity vector under the ECF. The Z axis points to the bottom of the sky, and the X axis is constrained by the direction of the fixing speed. The orbit prediction method is J4Perturbation. The suppression angles of the sun and the atmosphere-earth light of a star sensor are respectively 25° and 25°.

4.2. Output report

STK software provides a standard set of data reports to summarize key information, and all reports can be exported to popular spreadsheet tools in industry standard formats. In this paper, the required information includes the solar vector based on the satellite's ontology coordinate system, the time when the satellite enters and exits the illumination zone, and the duration of the illumination zone.

By using the vector geometry tool, I build a solar vector based on the satellite ontology coordinate system, which points from the centre of the sun to the centre of the satellite, and the size is the true distance from the satellite to the sun[9].The report tool is used to output the solar vector report of every moment in the simulation period under the satellite body coordinate system. The Sun item in the Report tool is used to calculate the visit to the Sun, and the time when the satellite enter and leave the illumination area and the duration in the illumination area are obtained.

4.3. Design of star sensor layout based on PSO

In practice, it is necessary to determine the orientation of the optical axis vector of the star sensor in the satellite body coordinate system and specify the specific installation position according to the specific satellite structure. Therefore, it is assumed that the starting point of the optical axis vector of the star sensor coincides with the centre of mass of the satellite body. The layout problem of star sensor is simplified to the problem of finding the best vertex coordinate of the optical axis vector in the satellite body coordinate system [10].

The starting point of the optical axis vector of the star sensor is coincident with the centre of mass of the satellite body. Taking the origin of the satellite body coordinate system as the centre, a three-dimensional sphere with the radius as the unit length is made. The vector formed from the origin to any point on the surface of the sphere is the apparent axis vector of the star sensor layout, forming a potential layout vector set. First, I remove the vector whose angle with the positive direction of the z-axis is less than the atmosphere-earth light suppression angle of the star sensor. Then the particle swarm optimization algorithm is used to calculate the mean value of the angle between the layout vector and the solar vector in the light time, and it is the optimal layout vector. Therefore, the star sensor layout problem is described as the problem of finding the global optimal solution in three-dimensional space.

4.3.1. Avoiding atmosphere-earth light interference. In the satellite body coordinate system. The conical plane, which takes the origin as the center of the circle, takes the positive direction of the z-axis as the central axis, and takes α_{se} as the semi-cone angle, intersects the unit sphere. Part of the spherical cone surface contains I area can be considered to be an accessible area of atmosphere-earth light [11].

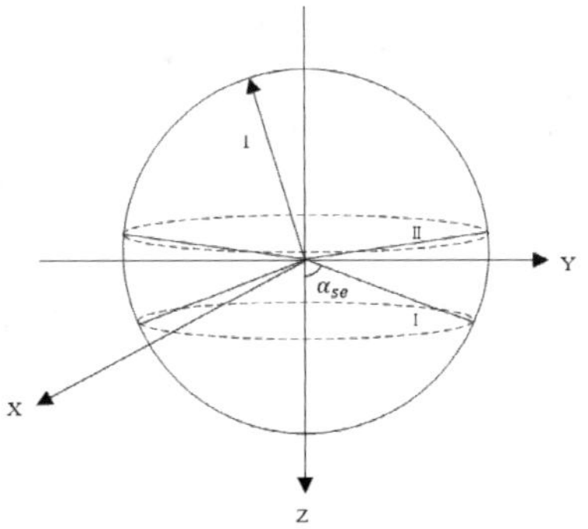

Figure 6. Unit sphere schematic diagram.

4.3.2. Obtaining the sun and moon exposure time of the satellite. Based on the Sun report from the STK output, I used MATLAB to filter out the set of sun vectors $\boldsymbol{b_j}$ under the sun's illumination time.

4.3.3. Using particle swarm optimization algorithm to obtain the optimal position. In order to reduce the interference of the direct sunlight on the star sensor, it is necessary to make the angle between the solar vector and the star sensor optical axis vector larger than the sun sensor suppression angle $\gamma_s$of the star sensor. According to the solar vector under the illumination time, the angle between the star sensor optical axis vector $\boldsymbol{a_i}$ and the vector $\boldsymbol{b_j}$ at a certain moment is:

$$\theta_{ij} = \arccos\left(\boldsymbol{a_i} \cdot \frac{\boldsymbol{b_j}}{\| \boldsymbol{a_i} \| \cdot \| \boldsymbol{b_j} \|}\right)$$

The coordinate points on the secure efficient sphere constitute the solution space of particle swarm optimization. The mean angle between the optical axis vector $\boldsymbol{a_i}$ of the star sensor and the solar vector $\boldsymbol{b_j}$ at a certain time in the period is taken as the evaluation index. Fitness function f of the algorithm is shown as follows, where k is the maximum moment in the cycle:

$$f = \frac{\sum_{j=0}^{j=k} arccos\,(\boldsymbol{a_i} \cdot \boldsymbol{b_j})}{k}$$

PSO initializes a group of random particles, namely random coordinate points on the safe and effective sphere. In each iteration, the position of the particle is updated by tracking the optimal solution found by the particle itself and the optimal solution found by the whole population. Calculate the best layout vector a_{i_best}=[-0.9280 -0.2984 0.2229], which is equivalent to a star sensor with an azimuth of -34° and an elevation angle of -40.5°. The layout orientation on the satellite is shown in the figure below.

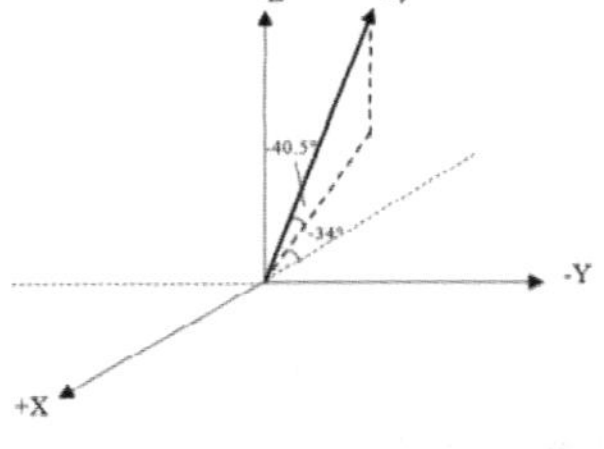

Figure 7. Star sensor installation angle diagram

4.4. Simulation Analysis of Incident Light Interference Based on STK

The Sensor1 is added into the scene created before. The sensor type is a simple cone. The pointing type is Fixed in Axes and the azimuth way is azimuth-elevation Angle（Az-El： -34°, -40.5°）. The reference frame is the satellite body coordinate system.

4.4.1. The semi-cone angle is set to 25°. And Line of Sight and Field of View are selected from the basic properties of the constraint. By using the output report function, the change of the angle between the apparent axis of the star sensor and the solar vector is calculated. As shown in the figure below, the minimum angle is 39.4°. At all times, the angle meets the condition of $\beta_{ss} \geq \gamma_s$, without interference from sunlight.

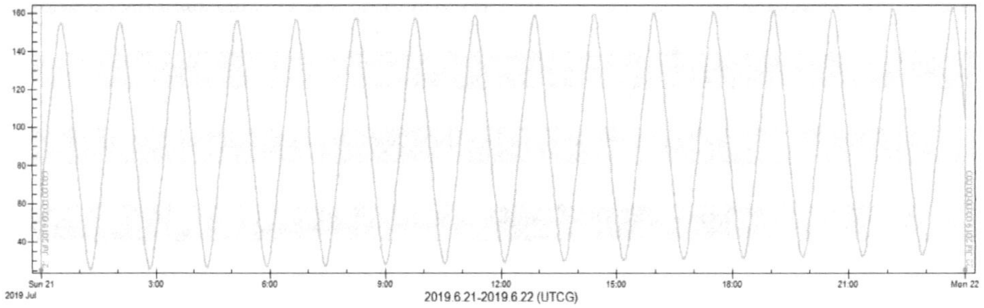

Figure 8. Change of the Angle between the apparent axis of the star sensor and the solar vector.

*4.4.2. The semi-cone angle is set to 25°.*Using the output report function, the change of the angle between the direction of the apparent axis and the earth vector is statistically calculated. and The Angle is always 130.5°. At all times, the angle satisfies the condition of $\beta_{se} \geq \gamma_e$. There is no earth-gas-light interference.

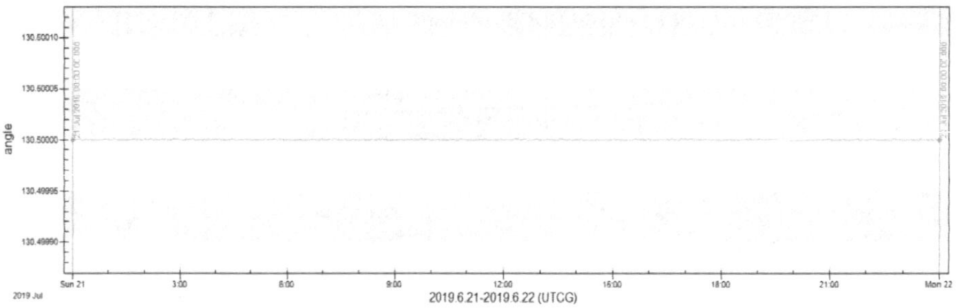

Figure 9. Change of Angle between the apparent axis of star sensor and the earth vector

5. Conclusion

In this paper, the model of incident light entering the viewing field of star sensor is established. Based on STK and PSO, the optimal position of star sensor is designed. The rationality and validity of this layout method are verified by STK. This method is applicable to the viewing field analysis of optical sensors on satellites of any orbital type. It can be used to design and optimize the placement of the star sensor on the spacecraft quickly and it meets the requirements of rationality and effectiveness of on-orbit work.

Reference

[1] CHEN C M, LI Y X, ZHOU W, GAO Y X. 2003. An Observation Study on the aviation Budget Characteristics for Earth atmospheric System. *Journal of Zhongshan University*. 42(1): 88-89

[2] AN M, ZHANG J, WANG X H and HAN B. 2013. Simulation and Analysis for the Stray Light Suppressing Based on STK. *Tactical missile technology*. (3): 59-66

[3] Li X Y, DU W and MO F. 2013. Simulation Analysis on View Field of Star Sensor Based on STK. *Spacecraft engineering* .22(2): 27-31

[4] ZHANG H， HAO Y J. 2013. Simulation and Analysis for the Stray Light Suppressing Based on STK. *Computer simulation*. 2(7): 83-86

[5] ZHAO LIN, SU Z H, HAO Y. 2013. Research on star sensor layout based on geometric position analysis. Transducer and Microsystem Technologies. 32(12): 34-37

[6] IAO Y, LIU G M and WEN Y L. 2006. Research on Modeling and Simulation for View Field of Star Sensor. *Journal of system simulation* .40(1): 38-40

[7] LIAO Z B, FU R M and ZHONG X Y. 2011. Design of specular baffle of star sensor. *Infrared and Laser Engineering*. 40(1): 66-69

[8] TANG Y, LU X and HAO C Y. 2004. Star Sensor Stray Suppression Analysis. *Aerospace Control.* 22(3): 58-61

[9] XIANG S H， ZHANG T. 2007. Calculation of Solar Direct Radiation on the Satellite External Surface Using STK. *Infrared Technology*. 29(9): 508-511

[10] XU K, CHEN Z G, ZHAO J H, DAI L and LI F. 2011. Layout design method of star sensor based on particle swarm optimization algorithm. *Journal of Jilin University （Engineering and Technology Edition）*. 49(3): 927-978

[11] YANG Y, WANG H L, LU J H. 2016. Analysis of Impact of Earth and Atmosphere Radiation on Star. *Optoelectronic Engineering*. 43(4): 8-14

Data Exchange of Digital Twins Based on AML in Space Science Experiment Equipment

Peng Guqi, Zhong Hongen

Technology and Engineering Center for Space Utilization, Chinese Academy of Sciences Beijing, China

E-mail: Pengguqi17@csu.ac.cn

Abstract. With the development of Computer Aided Design(CAD) and manufacturing technology, various industries are transforming to digitalization, and the realization of the digital model of aerospace products is an important part of "digital aerospace". Based on the theory of system engineering, this paper proposes a three-dimensional conceptual model of analysis of payload in the life-cycle of physical device design, as well as the concept of digital twin five- dimensional model in the design stage. Combined with the actual application requirements, this paper explores the content and standards of digital twin model construction of payloads, by using the model mapped to the AutomationML(AML) to realize the data exchange and fusion.

1. Introduction

With the advent of the Cyber Physical Systems, the improvement of network technology, information physics systems, digital technologies, and intelligent technologies promote the digitalization and intelligent transformation of aerospace products. Grieves(2014) proposed the Digital Twin concept to manage product information at product item level along its life-cycle. A Digital Twin appears as a virtual representation of the physical product, a digital shadow that reflects the structural, maintenance, performance, and operational health characteristics of the physical system. Its early application[4–7] is applied in the military and aerospace fields. NASA and Air Force Research Laboratories apply digital twin technology to aircraft health control [1, 2]. In 2012, the US Air Force Research Office proposed the concept of "vital digital twins" to emphasize the idea of a single source of real data. To date, in China there are many aerospace institutes have researched the digital twin model and more specifically, the model-based systems engineering(MBSE). Three Department has put forward MBSE-based aerospace product design solutions, and the 805 institution has proposed the application research of MBSE in the manned space field. In addition Advanced Space Academy proposed digital prototype engineering for the launch vehicle, etc[3].

Although digital technology has been widely applied in the development of products in the aerospace field, a few issues remain in the digital model research:

(i) Space Scientific Experimental Rack covers a wide range of fields (e.g. light, machine, electricity, heat, etc.), while the industry data exchange standards are not yet well defined, achieving inter-operability between fields is a significant challenge for digital twin development.

(ii) The development of Space Scientific Experimental Rack involves co-ordination between multiple research institutes based on the corresponding design management software for interdisciplinary fields. The above software is very mature but the interface compatibility between different software cannot be ignored.

(iii) The existing product data management only focuses on the technical state management. Such a way is dependent on the development and design documents, instead of realizing the digital model management.

Given the above situation, this paper has the aim to present a model architecture that offers services for different fields. We want to present a data exchange protocol using AutomationML for exchange data between the Digital Twin and other systems, and a methodology for communication for the data exchange.

2. Theoretical research on the construction of digital twin model

The digital twin model conforms to the basic characteristics: virtuality, uniqueness, multi- physical, multi-scale, hierarchical, integrated, dynamic, super-realistic, computable, multi- disciplinary[3]. For each physical entity, we can create a digital twin model of it using a modeling tool.

The American engineer Hall[8] has proposed the three-dimensional analysis model[9] of system engineering. We further propose a three-dimensional analytical conceptual model in the design stage, as shown in Figure 1 below.

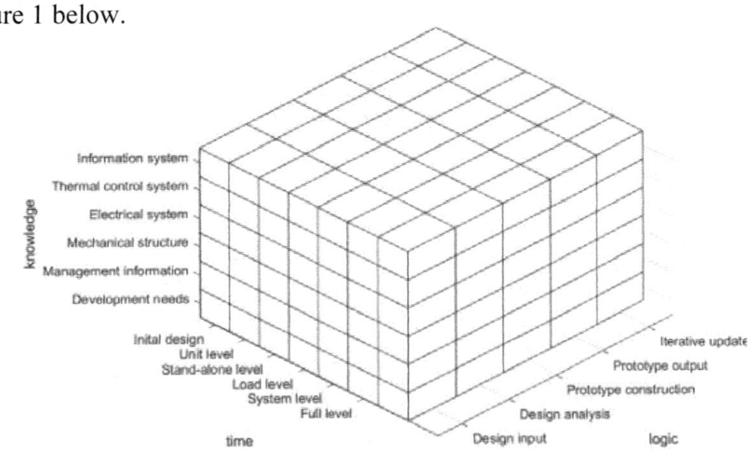

Figure 1. Three-dimensional analytical conceptual model.

In practice, the digital twin model is constructed based on time and knowledge slices. The life cycle of the model can be divided into six stages. Each level realizes further refinement of model attributes on the basis of the previous model. The preliminary design level mainly aims at the decomposition of data indicators and the unit level includes the subject domain model and uni-disciplinary model. The stand-alone level includes simulation model and realize part of data exchange work. The higher-level different model phases can meet different levels of usage requirements: load level should meet the basic test requirements, system level can realize information interaction with external system. Complete level includes data knowledge during the design phase test data and real-world operational data, which can be used for simulation prediction of its future operation, as well as support for design and simulation of new models. Therefore, the digital twin model is a dynamic digital representation of a physical world in its life-cycle.

According to the above modeling approach, the user has the possibility of creating a high level model of physical object. Different types of product models will be created during the different phases of product life-cycle, while creating a composite digital twin that integrates numerous parts from different fields is complicated, as shown in Fig.2.

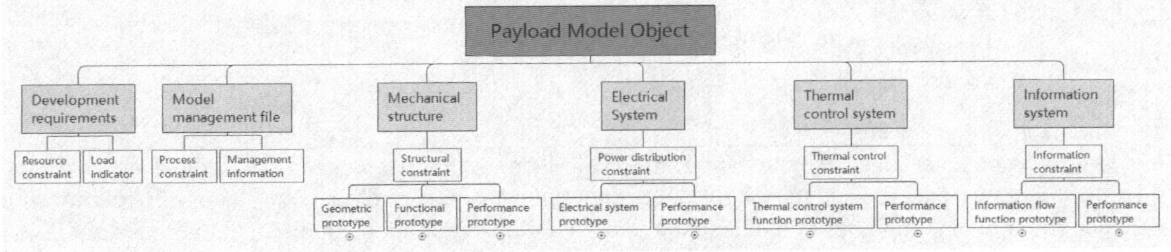

Figure 2. The composition of payload model object

To reflects the operational context of the physical twin, a digital twin model requires a physical twin for data acquisition and context-driven interaction. Furthermore, such models sometimes are inter-operable among each other. Having this in consideration, this paper proposes the concept of five-dimensional model of digital twin: $M_{DT} = (PE, VE, Ss, DD, CN)$. The most relevant topics for the creation of the digital twin can be summarized as follows[11]:

PE represents the physical entity. For example, a Space Scientific Experimental Rack payload is a simple system-level *PE*, including components, resource sets, and information collection, which enables resource co-ordination and information transmission between subsystems.

V E stands for virtual entity, $VE = (P_v, B_v, R_v, M_v)$, P_v is a physical model. B_v is a behavioral property model, describing the physical model behavior at different time and scale. R_v is mainly based on the accumulation of historical data, which continually updated with the latter's performance, maintenance, and health status data throughout the physical system's life- cycle. Mv mainly contains management information of virtual models, which can organize the storage model data along every stages.

Ss is a service model, which provides services to the physical device and services to the virtual model.

DD is a twin data model, which is the driver of each virtual device and service operation.

CN is the connection model, which is the key to constructing the digital twin system. It needs to ensure the connection between the physical entity and the virtual model, and the data transmission between the virtual models.

In summary, data associated with the physical system are supplied to the virtual environment to update the virtual model in the digital twin. Besides it is necessary to have a way of communication between the virtual models. To aggregate all the models, it needs to enable the exchange of each model[9, 10].

3. Data exchange of digital twin model

The existing modeling method is usually based on the corresponding interface of commercial software for interdisciplinary fields, such as ANSYS, ADAMS, MATLAB/Simulink. Therefore, the digital twin model involves the exchange and transfer of data between multiple model tools. To achieve the data exchange requirements of the above digital twin model, there are several ways to provide information to other systems. Through comparing various available open formats for model data exchange, as shown in Table1[14, 15], this paper proposes an AML-based data exchange approach, As shown in Fig. 3.

AML can be used to store and exchange engineering data in different engineering fields. It makes use of CAEX as a meta model for the storage and exchange of engineering models. By definition, CAEX supports object-oriented modeling for all aspects, which includes four concepts: role class, interface class, system unit class and instance level[12].

Table 1. Comparison of data content expressed by different languages

Feature	**Data Format**				
Mechanical Data	AML	STEP	JT	XPDL	XML
Electric Data	+	+	+	-	-
Process Control Data	+	+	-	-	-
Topological Information	+	-	-	+	-
Establishing Concept Relationships	+	+	+	-	+
Tracing Concept Dependencies	+	+	-	-	+
Mechanical Data	+	+	-	-	+

In order to organize an intermediate data transfer file format, it is necessary to define the organization form of the data, data type, data length, etc., and for each model object, the mapping relationship between the model data should be clarified[13]. The following is an example of a thermal simulation model under comsol (as shown in Fig. 4), which realizes the expression of model information by AML

IOP Conf. Series: Materials Science and Engineering **816** (2020) 012021 doi:10.1088/1757-899X/816/1/012021

and the data transfer interaction with the mechanical structure model. Different levels of digital twin model has inconsistent data exchange content, and the emphasis of model attributes is different. For a digital unit model at the overall unit level, the model under comsol contains the definition of global parameters, component information of the model, such as model geometry, model material, model physics Information, and simulation results of the model.

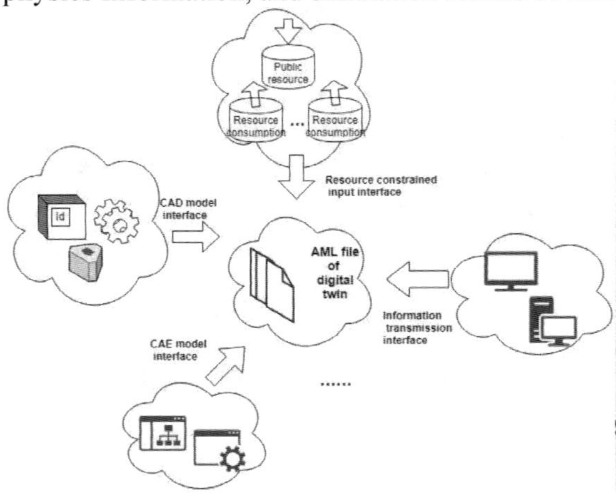

Figure 3. Data transfer in AML intermediate format.

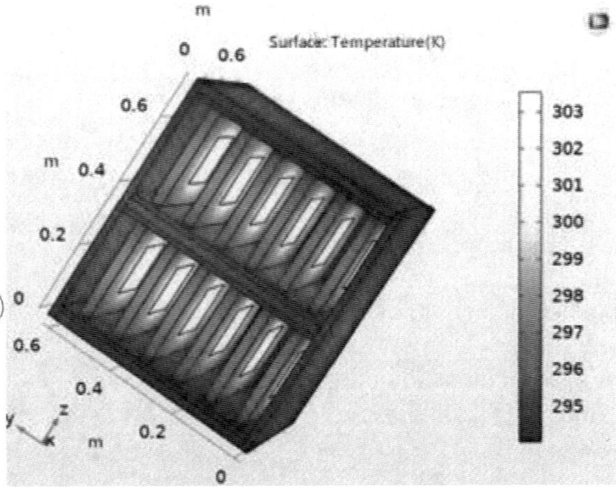

Figure 4. Thermal simulation model under comsol.

According to the above modeling framework, the electrical box is modeled. The model instance information and data element were mapped in AML format, and the relationship is shown in Table3. Different model data are mapped into four categories of AML, where the role class, interface class and system unit class are data category definitions in the model, and can be reused as the abstract part of the model. Instantiate the abstract attribute in the instance class to implement an instance of the electrical box heat transfer model.

The AML unit class is composed of management model, physical model and structural model data. The specific model data is represented by the attributes in the class. Each attribute has its content, format and other descriptions. System unit class mainly contains the cold plate of electrical box, heat transfer structure, heat source of the electrical box, box body, etc. In addition to the basic roles, the role class also has user-defined role classes such as components, structural parts, etc. The interface class also has basic interface class and user-defined interface classes, such as three different heat transfer interfaces. The heat simulation of model instantiation can be realized by applying the different attributes of the model, as shown in Fig.5. To better understand and view model attribute, the model can be extracted, and some parts are shown in Figure 6.

Table 2. Mapping relationships between model data in AML

Data element	Example	Mapping to AML
Hierachy level, Object	Heat transfer model module, Heat transfer system	Internal Element
Property	Structural attribute, Physical property	Attribute
Classification	Cooling device, Components, Structure	RoleClass
Simple Relation	Heat transfer	ExternalInterface with External Element for relation data
Directed Relation	Contact heat transfer	ExternalInterface with Internal Element for relation data
Groups	Heat source,cold plate, box body,electrical box	SystemUnitClass

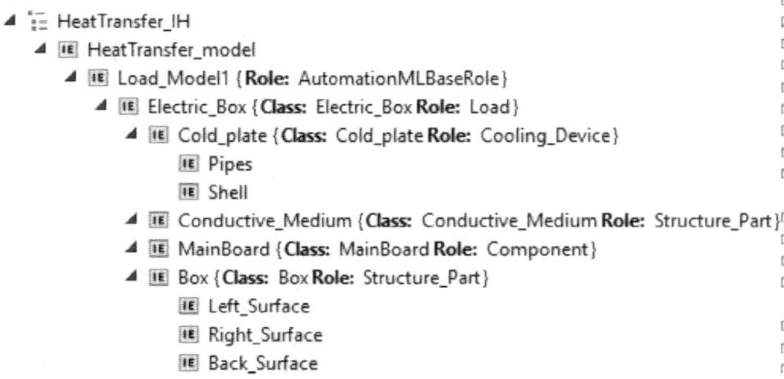

Figure 5. Data transfer in AML intermedi- ate format.

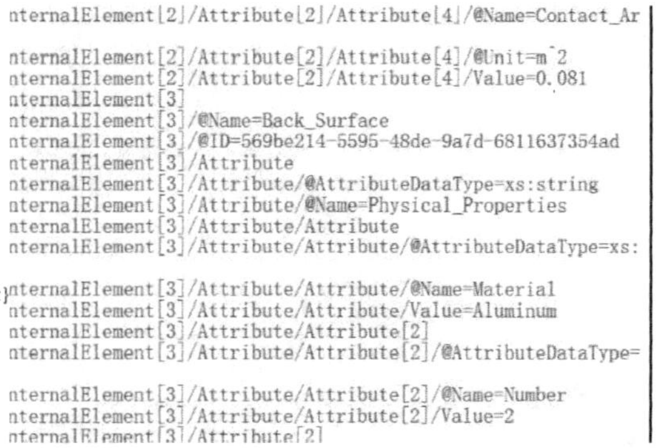

Figure 6. Thermal simulation model under comsol.

Through the above modeling method, the mechanical structure model of the typical electrical box model is created. For the mechanical structure, the data model lays particular emphasis on the related attribute information of the structural model. To ensure the understanding of the data content, the same attribute representation of the different domain models should be consistent. The AML-based model is shown in Figure 7.

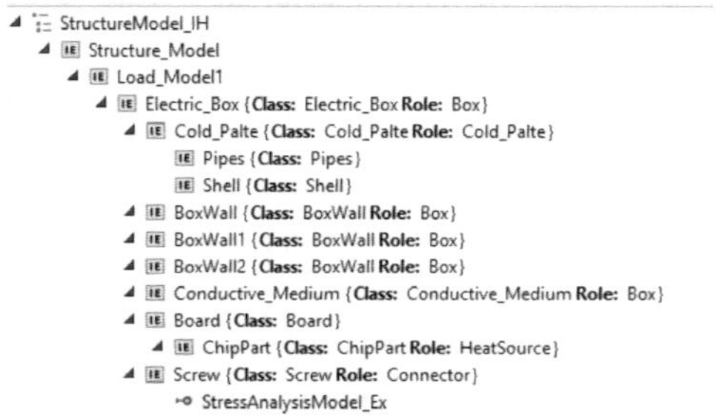

Figure 7. three-dimensional analytical conceptual model.

It can be seen that creating AML-based models can simplify the data exchange. The user has the possibility of representing physical components via a high-level model without having the professional knowledge in systems or programming languages and this model is used for the exchange of information, so that groups of proficiency in various fields and modeling tools can work together.

4. Future Works

This paper we proposed a methodology to create models using the AutomationML and an approach to exchange data between systems that making use of the Digital Twin. Through communication mechanism of these collaborative models, we can efficiently map the domain model to the specific AML-based models and they are shared to increase the consistency and collective knowledge around domain areas, so as to solve the compatibility problem of models and data exchange between the inter-systems and multi-systems.

At the same time, more and more industrial manufacturers are beginning to provide support for AML models, which stores engineering information and allows the modeling of physical and logical components in different aspects. Through the establishment of digital product profile, we can make it possible to track the performance and maintenance history of each physical twin over time, continuous research and development even after the sale of the physical twin. Specifically, these models can be exploited in upfront engineering, testing, system maintenance, and smart manufacturing.

5. Conclusion

In this work, we presented the main challenges of current Digital Twin technique and a approach to exchange data between fields using high level models. Through the theoretical research on the creation of digital twin model at the overall level, we propose a three-dimensional digital twin model in the design stage. The model structure of the payload is created according to the relevant field, and explained the model data connotation of the digital twin model of the payload under the five-dimensional model structure. To validate our proposal, we take thermal simulation as a case to study. From this model, we make use of the AutomationML models to exchange data between systems. The use of models allows that a user model a Digital Twin of the physical objects more efficiently. For future work, we intend to use the Digital Twin exchange model for other purposes.

6. References

[1] GLAESSGEN E, STARGEL D. The digital twin paradigm for future NASA and U.S. air force vehicles[C]// Proceedings of the 53rd AIAA/ASME/ASCE/AHS/ASC Structures, Structural Dynamics and Materials Conference, Reston, Va., USA: AIAA, 2012;7274-7260.
[2] TUEGEL E J, INGRAFFEA A R, Eason T G, et al. Reengineering aircraft structural life prediction using a digital twin[J]. International Journal of Aerospace Engineering, 2011,2011:1-14.
[3] Zhuang Cunbo, et al. (2017). Connotation, architecture and trends of product digital twin[J]. Computer Integrated Manufacturing Systems,2017,23(4):753-768.
[4] Tao Fei,Zhang Meng,Cheng Feng,etc.Digital twin workship: a new paradimg for future workshop[J].Computer Integrated Manufacturing Systems,2017,23(1):1-9.
[5] TAO F, ZHANG M, NEE A. Digital twin driven smart manufacturing[M]. Amsterdam, the Netherlands: Elsevier, 2019.
[6] TAO F, CHENG J, QI Q, et al. Digital twin-driven product design, manufacturing and service with big data[J]. The International Journal of Advanced Manufacturing Technology, 2018,94(9/10/11/12):3563-3576.
[7] TAO F, ZHANG M. Digital twin shop-floor: a new shop-floor paradigm towards smart manufacturing[J]. IEEE Access, 2017,5:20418-20427.
[8] Hall, A. D. (1962). A Methodology for Systems Engineering, Van Nostrand.
[9] Wang Xiaojun, Chen Haidong(2017).Launch Vehicle Digital Prototype Engineering, China Astronautic Publishing House.

IOP Conf. Series: Materials Science and Engineering **816** (2020) 012022 doi:10.1088/1757-899X/816/1/012022

Research on Flatness Rapid Measurement Technology of Large Deployable Antenna

LI Yun, ZHANG Jiabo, HAN Jianchao, CAI Zihui, WANG Yan, LI Yongliang

Beijing Spacecrafts, China Academy of Space Technology, 104 Youyi Road, Beijing, China

handanliyun@126.com

Abstract. The flatness of large deployable antenna needs to be measured several times during the deployment test, the theodolite measurement accuracy was low and the speed was slow, and the laser radar measurement range was limited. By building a flatness rapid measurement system for large deployable antenna, using autonomous navigation mobile technology to plan measuring path, and using digital photogrammetry technology to detect the flatness of antenna. It can realize rapid and automatic measurement for flatness of large antenna. The test results show that the accuracy of digital photogrammetry camera and laser tracker is no more than 0.05mm; the measurement accuracy of the flatness rapid measurement system shall not be more than 0.034mm away from the accuracy of laser radar, it can meet the requirements of flatness rapid measurement of large deployable antenna. This method can improve the efficiency and automation level of flatness measurement of large deployable antenna, and it lays a foundation for the plane measurement of large deployable antenna with larger size and more complex structure.

1. Introduction

The large deployable antenna is an important part of satellite. The planeness and pointing accuracy of the antenna after deployment directly affect the observation accuracy and performance of the satellite. Therefore, it is necessary to measure the planeness and pointing accuracy of the satellite antenna after deployment test [1-3].

At present, the satellite antenna is measured by theodolite or laser radar, the measurement accuracy of theodolite is low and its speed is slow; laser radar measurement range is limited, all of these have certain limitations for large deployment antenna. After the large deployment antenna is assembled, repeated deployment tests are required, the flatness needs to be measured for each deployment, therefore, the accuracy and efficiency of the existing measurement methods are low, which are not suitable for the flatness measurement of large-scale antenna [4-6].

In this paper, a fast measurement method of large deployable antenna plane is proposed, the panel of large antenna deployment system is measured automatically by photogrammetry technology; The autonomous navigation mobile system is used to realize the measuring path planning, improve the plane measurement efficiency and automation technology level of large antenna deployment system, It can meet the technical requirements of flatness rapid measurement of large antenna deployment system, and adapt to the development trend of large antenna in the future.

2. Construction of Flatness Rapid Measurement System for Large Deployable Antenna

The flatness rapid measurement system of large deployable antenna is shown in Figure 1. The system consists of autonomous navigation mobile system and high-precision digital photogrammetry system. In the measurement process of large deployable antenna, autonomous navigation mobile system and digital photogrammetry system can communicate in real time.

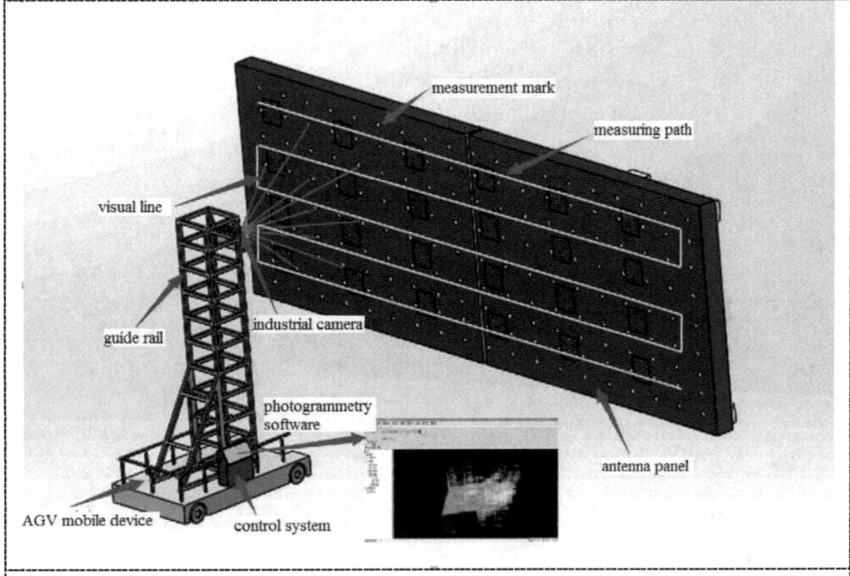

Figure 1. Flatness rapid measurement system of large deployable antenna

Autonomous navigation mobile system has optical automatic guidance function, it can make the digital camera move along the horizontal or vertical direction according to the planned path; the digital photogrammetry system is used to measure the flatness of large deployable antenna. After the autonomous navigation mobile system moves to the designated position according to the planned path, it sends signals to the digital photogrammetry system, the digital photogrammetry system takes pictures and feeds them back to the autonomous navigation mobile system after shooting, the autonomous navigation mobile system continues to move, and arrives to the next designated location for shooting, until the scanning of the whole antenna is completed. The closed-loop control of antenna measurement and autonomous movement is formed, so the automatic measurement of large antenna is realized.

2.1. Autonomous navigation mobile system

The autonomous navigation mobile system can move according to the planned path, and it can realize the movement of digital camera in horizontal or vertical direction, so the measurement efficiency is improved [7-9]. The autonomous navigation mobile system is mainly composed of autonomous navigation mobile device, guide rail, control system (industrial computer, controller, control software), etc. The autonomous navigation mobile device has the functions of omni-directional movement, intelligent obstacle avoidance, automatic positioning, etc. it can move or stop along the horizontal direction according to the specified path. The guide rail can be accurately positioned and can move or stop in the vertical direction. The main function of the control system is to receive the signal of arriving position for autonomous navigation mobile device or guide rail, convert it into control command to control measurement system for measurement, and receive the feedback signal of the measurement system, convert it into control command to control the autonomous navigation mobile device or guide rail to move according to the specified path.

IOP Conf. Series: Materials Science and Engineering **816** (2020) 012022 doi:10.1088/1757-899X/816/1/012022

2.2. Digital photogrammetry system

The digital photogrammetry system is composed of high-precision industrial camera, pan-tilt, photogrammetry software, control system, etc. The pan-tilt is a supporting device for installing and fixing cameras, which is suitable for scanning in a wide range, and it can expand the field of view of the camera. The photogrammetry software processes the collected image data, and completes the flatness analysis and other related functions. By using the high-resolution measurement camera, more than two digital images of the measured object are acquired at different positions and directions. After image scanning processing, sign recognition, image matching, spatial triangle intersection and beam adjustment, the three-dimensional coordinates of the object to be measured are obtained. After processing, 3D data of tested object will be entered into the coordinate system, it calculates the planeness and pointing accuracy of large antenna [10-12].

3. Path Planning of Measurement System

In order to have a better measurement network type and the angle of intersection, the measuring method of 45° angle between the main axis of visual line and the normal line of the antenna panel is adopted. During the measurement, first place the camera horizontally for shooting measurement, then rotate the camera 90° along the main axis of visual line to measure the same area. Move the autonomous navigation mobile device to the next location to continue the measurement, until the entire antenna panel area is measured. The measured attitude of the camera is shown in figure 2; the camera is placed horizontally, with the main axis of camera visual line at an angle of 45° from the normal of antenna panel.

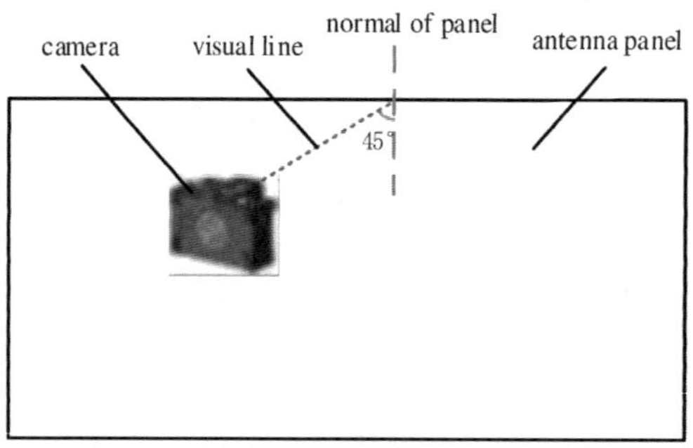

Figure 2. Measurement attitude of camera

The distance between the camera and the antenna panel is 1.4*m*, and the camera parameter field of view is 72°×51°. When the camera is placed horizontally, the effective observation range of camera is 2.2*m* in the horizontal direction, and 1.2*m* in the vertical direction. According to the measurement method of 60% overlap degree in both horizontal and vertical directions, the moving distance in each horizontal direction is 0.85*m*, and that in the vertical direction is 0.45*m* after a row of shooting is completed.

When the camera rotates 90° along the main axis of visual line, and is placed vertically for measurement, the effective observation range of camera is 1.9*m* in the horizontal direction, and 1.6*m* in the vertical direction. According to the measurement method of 60% overlap degree in both horizontal and vertical directions, the moving distance in each horizontal direction is 0.75*m*, and that in the vertical direction is 0.55*m* after the row is finished.

In order to ensure that the overlap of the camera's horizontal and vertical placement can reach 60%, so the camera's moving distance is 0.75*m* in the horizontal direction and 0.45*m* in the vertical direction. The measurement range of antenna panel is 10*m* × 2.5*m*. The camera needs to move 14 times in the horizontal direction and 6 times in the vertical direction.

4. Accuracy verification of measurement system

4.1. Accuracy verification of digital photogrammetry system

The accuracy of digital photogrammetry camera is detected with a laser tracker (as shown in Figure 3). A digital photogrammetric field is established, in which six target balls are placed as measurement points. By comparing the measurement results of laser tracker and digital photogrammetry camera, the accuracy of measurement camera is verified.

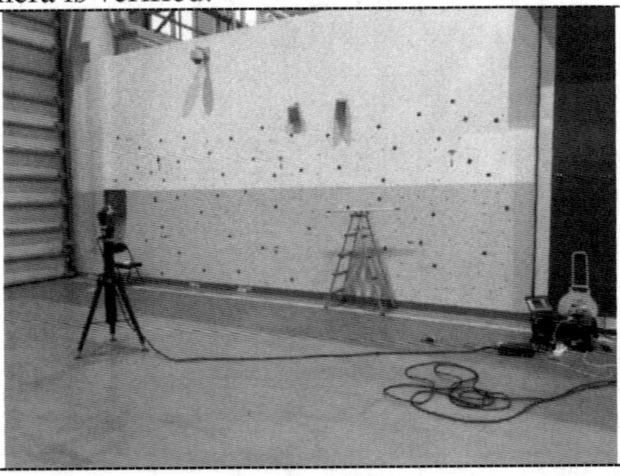

Figure 3. The accuracy verification test of camera

The difference between the digital photogrammetric camera and the laser tracker is shown in Table 1. It can be seen from Table 3 that the accuracy deviation between the digital photogrammetric camera and the laser tracker is not more than 0.05*mm*, which meets the accuracy requirements.

Table 1. Contrast of measured result

Target ball	Coordinate X (*mm*)	Coordinate Y (*mm*)	Coordinate Z (*mm*)	Mean square deviation(*mm*)
1	-0.02	0.03	0.02	0.041
2	0.02	-0.01	0.01	0.025
3	-0.03	0.02	-0.02	0.041
4	0.01	-0.03	0.01	0.033
5	0.02	0.02	-0.01	0.03
6	0.01	-0.01	-0.01	0.017

4.2. Experimental verification of flatness rapid measurement system

Verify the principle of flatness rapid measurement system (as shown in Figure 4), Paste photogrammetric marks, including coding point marks and single point marks, within $10m \times 2.5m$ of the flat wall, and place the measuring datum ruler. According to the measuring path planning, set the distance between the horizontal movement of autonomous navigation mobile device and the vertical movement of guide rail, run the flatness rapid measurement system to measure, and carry out five times measurements on the measurement area. The measurement results are shown in Table 2. Then the laser radar is used to measure once. Through data comparison, it can be seen that the maximum deviation between the accuracy of the plane measurement system and the laser radar is 0.034*mm*, and the minimum deviation is 0.009*mm*, which meets the use requirements.

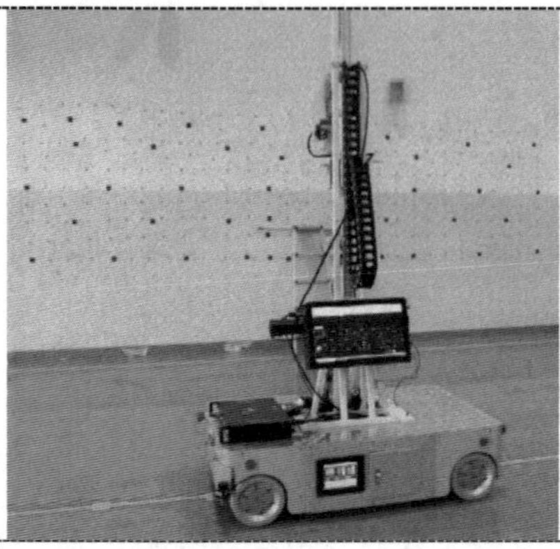

Figure 4. The Principle verification test of flatness measurement system

Table 2. Measurement result

Number	Measurement result of flatness measurement system (*mm*)	Measurement result of laser radar (*mm*)	Measurement deviation (*mm*)
1	5.871		0.021
2	5.884		0.034
3	5.841	5.85	0.009
4	5.824		0.026
5	5.836		0.014

5. Conclusion

The rapid measurement technology of large deployable antenna can realize the flatness measurement of large deployable antenna. The antenna panel is measured by digital photogrammetry technology, the accuracy deviation between digital photogrammetry camera and laser tracker is not more than 0.05*mm*; the measurement accuracy deviation between the flatness rapid measurement system and the laser radar is not more than 0.034*mm*, which can meet the requirements of flatness rapid measurement of large deployable antenna. This technology solves the problems of low precision and slow speed of theodolite measurement system, and solves the problem limited measurement range of laser radar measurement system. It improves the measurement efficiency and automation level of large deployable antenna, and lays a foundation for plane measurement of large deployable antenna with larger size and more complex structure.

References

[1] Macdonald M, Mcinnes C, Hughes G. Technology Requirements of Exploration Beyond Neptune by Solar Sail Propulsion [J]. Journal of Spacecraft and Rockets, 2010, 47(3): 472-483.

[2] Tian Dake. Design and experimental research on truss structure for modular space deployable antenna [D]. Haerbin: Harbin Institute of University, 2011.

[3] Cui N G, Liu J F, Rong S Y. Solar Sail Spacecraft Dy-namic Modeling and Solving [J]. Acta Aeronautica, 2010, 31(8): 1565-1571.

[4] Christopher G M, William G D, Michael L P, et al. Experimental Characterization and Finite Element Analysis of Inflated Fabric Beams[J]. Construction and Building Materials, 2009, 23(5): 2027-2034.

[5] Meguro A, Ishikawa H, Tsujihata A. Study on Ground Verification for Large Deployable Modular Struc-tures[J]. Journal of spacecraft and rockets, 2006, 43(4): 780-787.

[6] Zhang Xueyu. Single AGV Path Planning Based on Automated Logistics [D]. Guangzhou: Guangdong University of Technology of, 2016.

[7] Chen Liangliang. Indoor mobile robot autonomous navigation and grasing planning [D]. Guangzhou: Guangdong University of Technology of, 2016.
[8] Huang Guiping. Study on the Key Technologies of Digital Close Range Industrial Photogrammetry and Appfications [D]. Tianjin: Tianjin University, 2005.
[9] LI Yun, ZHANG Jiabo, HAN Jianchao, CAI Zihui, LU Ligang, ZHAO Linna. Assembly Measurement and Real-time Feedback Adjustment Technology of Large Antenna [J]. Journal of Nanjing University of Aeronautics and Astronautics, 2019, 51(S):88-93.
[10] ZHANG Xinpan. Application of Digital Industrial Photogrammetry to Realize Antenna Reflector Pose Recovery [J]. Journal of Astronautic Metrology and Measurement, 2017, 37(6):16-21.
[11] LI Zhanli, LIU Mei, SUN Yu. Research on calculation method for the projection of circular target center in photogrammetry [J]. Chinese Journal of Scientific In-strument, 2011, 32(10): 2235-2241.
[12] Fan Shenghong. Research on the key thchnologies of track slab quick detection based on digitial photo-grammetry [D]. Beijing: China University of Mining and Technology, 2014.

The Interference Analysis of the Terrestrial Components of International Mobile Telecommunications to the Satellite Mobile Communication System in S-band

Li Ying[1], **Li Yan**[2]

[1]Beijing Institute of Tracking and Telecommunications Technology
[2]Chinese People's Liberation Army No.93642

[1]liying@bittt.cn; [2]hbwzxwd@sina.com

Abstract. In accordance with the Radio Regulations, the S-band (Earth-to-Space 1980-2010 MHz, Space-to-Earth 2170-2200 MHz) is currently a shared frequency band both for the mobile service and the mobile-satellite (MSS) service. Both the terrestrial and satellite components of IMT have already been deployed or are being considered for deployment in some countries with the frequency bands 1 980-2 010 MHz and 2 170-2 200 MHz. Both the terrestrial and satellite components of IMT have already been deployed or are being considered for deployment. When Therefore, the potential interference issues between the two systems need to be studied. According to the characteristics of the terrestrial components of IMT and satellite mobile communication system, this paper analyzes the impact of the terrestrial components of IMT on satellite mobile communication system, and gives measures to eliminate interference.

1. Introduction

In accordance with the Radio Regulations[1], the S-band (Earth-to-Space 1980-2010 MHz, Space-to-Earth 2170-2200 MHz) have been identified for International Mobile Telecommunications（IMT）, which is currently a shared frequency band both for the mobile service and the mobile-satellite (MSS) service. At present, some countries, including China, have deployed the mobile-satellite systems within this frequency band. At the same time, the terrestrial components of IMT has developed to the fourth generation (4G), and gradually advanced to the fifth generation (5G). With the rapid development of the terrestrial components of IMT, the demand for frequency is also increasing. Some countries are planning to deploy the S-band terrestrial components of IMT. If the countries covered by satellite beams deploy the terrestrial components of IMT, it will have an impact on the MSS. Therefore, the potential interference issues need to be studied.

This paper first introduces the interference scenario, and then gives the typical parameters of the MSS and the terrestrial components of IMT. Based on the typical parameters, the interference from the terrestrial components of IMT to the MSS uplink and downlink are analyzed. Finally, measures and suggestions to reduce interference are concluded.

2. Interference Scenarios

The terrestrial components of IMT interferes with MSS satellite systems mainly in A1 and A2 scenarios (shown in Figure 1). This paper focuses on the A1 and A2 scenarios.

Scenario A1: the uplink interference issue from UEs and BSs to the satellite in 1 980-2 010 MHz

Scenario A2: the downlink interference issue from BSs to satellite MES in 2 170-2 200 MHz

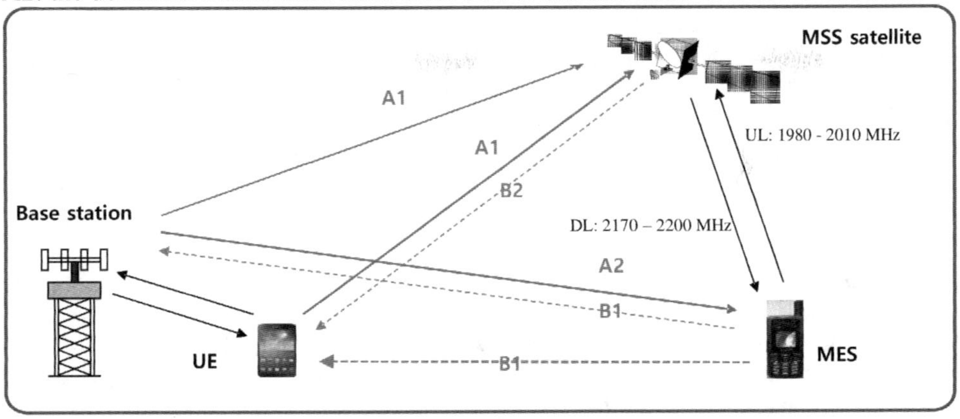

Figure. 1. Interference scenarios between the MSS and terrestrial components of IMT

IMT: International Mobile Telecommunications
UE: User Equipment of the terrestrial components of IMT
BS: Base Station of the terrestrial components of IMT
MES: Mobile Earth Station of the MSS
GSO: Geo-Stationary Orbit

3. MSS and Terrestrial Components of IMT Typical Parameters

Table 1 and Table 2 below list the receive parameters respectively of the GSO MSS[2] satellite and MES.

Table1. GSO MSS satellite receive parameters

Type	Value
Rx antenna gain (dBi)	41.6
Satellite G/T (dB/K)	20
Polarization	Circular
Receiver noise temp (K)	575
Minimum elevation angle from the MES	5°
Bandwidth (kHz)	21.6 to 1 296
Receiver thermal noise (dB(W/Hz))	–201
Interference criteria for purposes of this study	ΔT/T = 6% in-band

Table 2. MSS System MES characteristics

MES type	Maximum antenna gain	Maximum e.i.r.p. density	Receiving system noise temp.
3G Handset	0 dBi	–7 dBW/21.6 kHz	290K
Portable	2 dBi	5 dBW/21.6 kHz	200 K
Vehicular	10 dBi	20 dBW/216 kHz	250 K

Table3. The terrestrial components of IMT parameters

	UE	BS
Interference e.i.r.p. psd (dBmW/Hz)	-36.6	-15.3（main-lobe） -35.3（side-lobe）

IOP Conf. Series: Materials Science and Engineering **816** (2020) 012023 doi:10.1088/1757-899X/816/1/012023

4. Analysis

4.1. The uplink interference analysis from UEs and BSs to the satellite (Scenario A1)

Currently, a recommended value for the protection criterion related to interference from IMT terrestrial to MSS does not exist.

According to Appendix 8 of the Radio Regulations, the geostationary orbit satellite long-term interference protection criterion for the satellite service is in (1). So we use I/N= − 12.2 dB as the protection criterion during the study.

$$I/N=-12.2 \text{ dB} \quad (1)$$

The maximum EIRP density of the IMT terrestrial emission is calculated in (2).

$$\text{EIRP density}_{tmax}=-12.2+Ls-Gr+KT \quad (2)$$

The interference analysis is showed in Table 4. As can be seen from the table, when the aggregate transmit power spectral density of the UEs and BSs does not exceed -64.5 dBW/Hz, it will not interfere with the reception of the satellite repeater.

It can be showed from the terrestrial component of IMT parameters(Table 3) that when the elevation angle of the BS is high, that is, when the main lobe of the BS directly faces the satellite, it will interfere with the reception of the satellite repeater. When the side lobe is facing the satellite, it may not interfere with the reception of the satellite repeater. However, if the satellite receives multiple BS and UE signals in the same receiving beam, even if the main lobe of BS is not directly faced with the satellite, it will also interfere with the satellite repeater. The spatial distribution of the BSs and UEs belonging to the terrestrial IMT deployment(s) in view of the satellite will impact the e.i.r.p. of the Earth to space interference emissions predominately as a function of elevation angle to the satellite.

Table4. Interference analysis of the Scenario A1

Type	Value
Frequency(GHz)	1.995
I/N（dB）	-12.2
Distance(km)	36000
Free Space Loss(dB)	189.6
Sat Rx Gain(dBi)	41.6
Sat Noise Temp(K)	575
IMT Terminal aggregate EIRP density(dBW/Hz) max	-65.2

Furthermore, the position of the BSs and UEs with respect to the satellite beam coverage area will impact the received interference power at the satellite as a function of satellite beam directivity. However, it has been observed in the aggregate interference analysis results that terrestrial IMT interference from the BS terminals to satellite transponders is very serious. Therefore, it is recommended that the BS transmitting frequency band uses 2170-2200MHz to reduce interference to the satellite uplink.

The aggregate interference from UEs into GSO satellite does not show exceedance of the I/N protection criterion of -12.2dB. The margin is in the range of 1.4 dB to 26.1 dB depending on the location, the elevation angle of the satellite relative to the locations of the UEs, as well as the geographical area in which the aggregate interference was calculated.

Based on the above mentioned it is could be concluded that the aggregate interference from UE IMT stations towards GSO satellite system is not expected. At the same time the aggregate interference from BS IMT stations towards GSO satellite system is expected.

As can be seen from the above summary results, the risk of interference from IMT base station deployment is very high, while there is less risk of interference from IMT UEs.

4.2. The downlink interference analysis from BSs to satellite MES in 2170-2200 MHz(Scenario A2)

The signal transmitted by the BSs in the 2170-2200MHz frequency band will interfere with the reception of the MES. So the minimum separation distances should be calculated.

The interference analysis methodology is based on the propagation model from Recommendation ITU-R P.452[3] and taking into account actual antenna gains in the direction of interference.
Several parameters are showed in the following Table V. Path loss versus distance plots are obtained to calculate separation distances between IMT systems and MES at frequency 2185 MHz, assuming a 100% land path. For a given value of required link loss, the corresponding separation distance may be determined from the graphs. These graphs are used to determine the separation distances for scenario A2.
From this propagation loss the resulting separation distances were obtained for several IMT BSs with various e.i.r.p. levels to predict the interference into MESs over and paths. A static analysis was done for propagation losses not exceeded for time percentages of 10% and 50% over 100 % land path without clutter effect.
Given a lack of protection criterion for compatibility studies between the terrestrial and satellite components of IMT, considers the most appropriate criteria to be used for studies is 6% (-12.2 dB I/N). The results of the studies undertaken are summarized below:
The minimum separation distance between a single BS and MES of the systems studied over a 100% land path using I/N protection criterion of -12.2 dB varied in different studies:

- For p=10%, studies had results that varied from 48 to 123 km;
- For p=50%, studies had results that varied from 35 to 51 km.

The potential interference may be mitigated by one or more of: assessment of terrain and clutter effects and system characteristics, deployment environments, and separation distance.

Table 5. Parameters values for land path model

Land path model parameters	Input values	Remark
f (GHz)	2.185	
p (%)	10/50	
htg (m)	1.5	MES
hrg (m)	1.5/30	UE/BS
Gt (dBi)	2	MES portable
Gr (dBi)	18	BS rural
ΔN (N-units/km)	41.13	
N0 (N-units)	320.5	
dct (km)	500	
dcr (km)	500	
Path profile height	0	
Path profile zone	A2	Inland

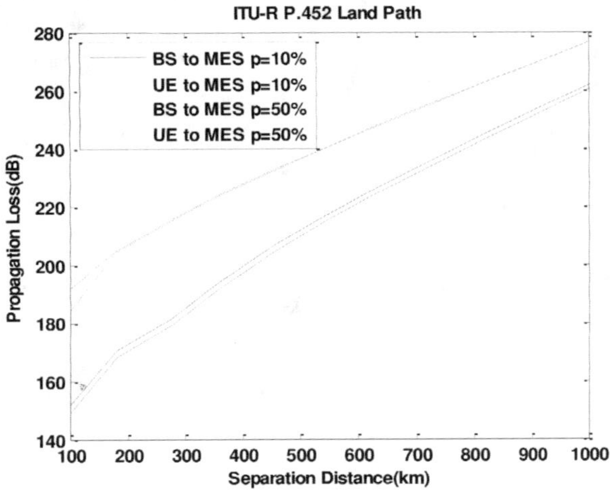

Figure2. Propagation loss of a land path from a IMT System to a MES

IOP Conf. Series: Materials Science and Engineering **816** (2020) 012023 doi:10.1088/1757-899X/816/1/012023

5. Interference Avoidance Measures

In the frequency band 1 980-2 010 MHz, it was observed that the level of potential interference from IMT BS into MSS is high, while the level of potential interference from IMT UE into MSS is low. The spatial distribution of the BSs and UEs belonging to the terrestrial IMT deployment(s) in view of the satellite will impact the e.i.r.p. of the Earth to space interference emissions predominately as a function of elevation angle to the satellite. It is recommended that the BSs use the 1980-2010MHz frequency band to work in the receiving mode.
In the frequency band 2170-2200 MHz, co-frequency deployment of independent MSS and terrestrial components of IMT in the same geographical area is not feasible. It is recommended that takes some techniques, such as the use of an appropriate guardband or the unified design of the two systems to ensure coexistence and compatibility between the terrestrial and satellite components of IMT. The implementation of mitigation measures may be considered on a case by case basis by administrations.

6. Conclusions

For areas where the terrestrial components of IMT cannot be deployed or the deployment cost is too high, the satellite mobile communication system can meet the needs of user mobile communication and global coverage with no blind zone. In addition, MSS are an irreplaceable under emergency communications such as during the natural disasters or wartimes. China and some other countries has deployed or planned to deploy the S-band MSS. However, the S-band shared by the terrestrial components of IMT and MSS will cause mutual interference. Studies to evaluate the co-existence and compatibility of terrestrial and satellite components of IMT are very important. It can be concluded that independent MSS and the terrestrial components of IMT cannot coexistence. Administrations planning satellite IMT systems should firstly consider possible technical and operational measures on satellite IMT systems, because they are implementable by the efforts of satellite IMT only. This paper provides some references for the design and construction of the MSS.

7. References

[1] The Radio Regulations,2016
[2] ITU-R S.672-4, Satellite antenna radiation pattern for use as a design objective in the fixed-satellite service employing geostationary satellites.
[3] ITU-R P452-16, Prediction procedure for the evaluation of interference between stations on the surface of the Earth at frequencies above about 0.1 GHz, 2015.07.
[4] Reference earth-station radiation pattern for use in coordination and interference assessment in the frequency range from 2 to about 30GHz [M], 1993, Rec.ITU-R S.465-5
[5] Radiation diagrams for use as design objectives for antennas of earth stations operating with geostationary satellites [M],2003, Rec.ITU-R S.580-6
[6] Method of calculation for determining if coordination is required between geostationary-satellite networks sharing the same frequency bands [M], the Radio Regulations II APPENDIX 8, 2016, 125-131
[7] Dennis Robby, Satellite Communications[M]. Zheng Baoyu, translated, Beijing: China Machine Press,2011:234-236
[8] Tan Weizhi, Hu Jingang. Spacecraft systems engineering[M]. Beijing: China Science and Technology Press, 2009:334-336.

AUTHOR INDEX